T0339908

UNCERTAINTIES IN GPS POSITIONING

UNCERTAINTIES IN GPS POSITIONING

A Mathematical Discourse

ALAN OXLEY
Bahrain Polytechnic
Kingdom of Bahrain

ACADEMIC PRESS

An imprint of Elsevier
elsevier.com

Academic Press is an imprint of Elsevier
125 London Wall, London EC2Y 5AS, United Kingdom
525 B Street, Suite 1800, San Diego, CA 92101-4495, United States
50 Hampshire Street, 5th Floor, Cambridge, MA 02139, United States
The Boulevard, Langford Lane, Kidlington, Oxford OX5 1GB, United Kingdom

Notices
Knowledge and best practice in this field are constantly changing. As new research and experience
broaden our understanding, changes in research methods, professional practices, or medical treatment
may become necessary.

Practitioners and researchers must always rely on their own experience and knowledge in evaluating
and using any information, methods, compounds, or experiments described herein. In using such
information or methods they should be mindful of their own safety and the safety of others, including
parties for whom they have a professional responsibility.

To the fullest extent of the law, neither the Publisher nor the authors, contributors, or editors, assume
any liability for any injury and/or damage to persons or property as a matter of products liability,
negligence or otherwise, or from any use or operation of any methods, products, instructions, or ideas
contained in the material herein.

Library of Congress Cataloging-in-Publication Data
A catalog record for this book is available from the Library of Congress

British Library Cataloguing-in-Publication Data
A catalogue record for this book is available from the British Library

ISBN: 978-0-12-809594-2

For information on all Academic Press publications
visit our website at https://www.elsevier.com/

**Working together
to grow libraries in
developing countries**

www.elsevier.com • www.bookaid.org

Publisher: Glyn Jones
Acquisition Editor: Glyn Jones
Editorial Project Manager: Charlotte Cockle
Production Project Manager: Poulouse Joseph
Cover Designer: Matthew Limbert

Typeset by SPi Global, India

CONTENTS

LIST OF FIGURES

Appendix A

Appendix E

LIST OF TABLES

PREFACE

Navigation involves monitoring and controlling the movement of a craft or vehicle from one place to another. Position fixing is a branch of navigation concerned with determining the position of an aircraft, person, or ship on the surface of the Earth. Many position fixing technologies involve the processing of radio signals.

There are space-based navigation systems and terrestrial-based navigation systems. The Global Positioning System (GPS) is an example of the former. An example of a terrestrial system is the position fixing of a mobile phone by analyzing radio signals between (several) radio towers of the network and the phone.

We are all familiar with the fundamentals of GPS receiver: It is a device that stores a map and receives signals from a number of satellites enabling the receiver to determine its location. This is often done in near real-time. *Uncertainties in GPS Positioning: A Mathematical Discourse* describes the calculations performed by a GPS receiver and describes the problem associated with making sure that the estimated location is in close agreement with the actual location. Inaccuracies in estimating a location could have serious consequences.

The intended readership is individuals interested in GPS, such as university students. The reader will benefit from being able to: understand how a GPS receiver calculates its position; understand why the calculated position is only an approximation to the true position; gain some appreciation of the factors which contribute to the difficulties in calculating an approximation of the true position; gain some appreciation of the mathematical steps that are employed in order to reduce errors in the approximation.

The impetus for the book came from the fact there is insufficient literature describing the calculations performed by a GPS receiver and little detailing the accuracy of the calculations.

The book comprises the following chapters:

Chapter 1—Positioning and Navigation Systems

Chapter 2—Introduction to GPS

The rudiments of what GPS is from a user's perspective.

Chapter 3—Basic GPS Principles

This chapter describes how the shape of the earth is approximated in the mathematical calculations, the satellites that make up the GPS system, the axis system that is commonly used, the type of clock used by a satellite and the type of clock used by a receiver, the mathematical notation used (for denoting distance from a receiver to a satellite, the speed of light, a pseudodistance, etc.), and the physical.

The physical phenomena that can affect the distance calculation.

Chapter 4—Signals from Satellites to Receiver—GPS

Chapter 5—GPS Modernization

Chapter 6—Signals from Satellites to Receiver—Other Satellite Navigation Systems

Chapter 7—Solution of an Idealized Problem

This chapter describes how a receiver position can be calculated accurately given the exact position of at least four satellites and the exact pseudodistances; the complete calculation for example data; and shows actual data that is available and from which a receiver must calculate its location.

Chapter 8—Sources of Inaccuracy

This chapter describes the problem of a receiver inaccurately calculating a location, explains why a satellite's position is not known precisely, explains why a pseudodistance is not known precisely, explains the sources of inaccuracy, and also describes how the various factors that figure in the calculations depend on one another.

Chapter 9—Learning from Experience

Chapter 10—Error Distribution in Data

This chapter describes what can be said about how the errors are distributed for each item of data—a satellite location, a pseudodistance, etc., and gives examples of probability distributions for satellite location, pseudodistance, etc.

Chapter 11—Improving Accuracy with GPS Augmentation

Chapter 12—GPS Disciplined Oscillators

GPS positioning is used in numerous applications and so the overall topic of GPS is multidisciplinary covering diverse topics such as signal processing and computing. An everyday application is a GPS car navigator. There are also GPS-enabled smartphones. Having the ubiquitous facility to easily determining one's position has spawned context-aware services, where location is part of the context. (Sometimes context-aware services are referred to as location-based services.) The act of finding the location of a retail outlet, bank branch etc., in a new city has been transformed.

On the horizon, is the development of GPS such that it can be used in environments that have hitherto been unsuited to it, such as indoors. This will herald even more ideas for applications.

The fundamentals of how a positioning system derives a user's position are essentially common, whether the system is GPS or some other system. This commonality aids the integration of systems.

GPS positioning is based on calculations performed from signals between a user receiver and the satellite's that are in view. Accuracy can be improved with the addition of measurements derived from other sources. In these cases, we say that GPS has been "augmented." Example augmented systems include assisted GNSS, differential GPS, EGNOS, and RTK networks.

Techniques exist to handle very weak radio signals, such as GPS signals received indoors. Techniques also exist for combining GPS-derived measurements with those derived from sensors. It is also possible for many GPS receivers to interact with the aim of improving positioning accuracy.

The book contains a large number of terms which the author has endeavored to explain. Some of these are well known, such as GIS. (A geographic information system (GIS) or geospatial information system is a system that stores spatial or geographical data.) Others are less well known, such as an estimator. (An estimator is a rule for calculating an estimate of a given quantity based on observed data. An estimator is a statistic (ie, a function of the data) that is used to infer the value of an unknown parameter.)

CHAPTER 1

Positioning and Navigation Systems

Certain new applications of IT change our daily lives. There are a growing number of such applications; correspondingly, more people are opting to use them. Examples include fitness applications, Global Positioning System (GPS) tracking, health monitoring, and navigation systems. In the 1970s, only a science fiction writer could have foresaw that people would be allowing an application to access a little of their personal data, their geolocation, in order to use a service.

NAVIGATION

Navigation has long been of interest to mathematicians.

History of Navigation

Historically the astrolabe was used to calculate the positions of the sun and stars, among other things. Instructions for building an astrolabe are given by St. John's College, University of Cambridge (2014). It is generally accepted that the astrolabe was invited in either the first or second century BC. The odometer is an instrument for measuring distance traveled. The historical text of the Song Shi, recording the people and events of the Chinese Song Dynasty, gave a detailed description of the odometer. Historically the pole stars have been used for navigation as, throughout the night, they are visible above the horizon and their apparent positions remain virtually fixed. The Minoan sailors of Crete are an early example of those that have used celestial navigation. The position of the pole stars changes over time due to the wobble of the Earth on its axis. Another aid to sailors was nautical charts. The earliest known world maps date from the fifth and sixth century BC. Ptolemy (AD 100–170) was a key figure in early map making. Sailing directions have existed since before the time of Ptolemy.

Uncertainties in GPS Positioning
http://dx.doi.org/10.1016/B978-0-12-809594-2.00001-0

In 1884 the Greenwich meridian was adopted as the prime meridian of the world (i.e., 0 degrees longitude). This is of fundamental importance to positioning.

Type of Navigation

There is a type of navigation where readings are taken and retaken at regular intervals. Satellite-based navigation makes use of this approach. There is a type of navigation where several receivers are used to detect the position of an object of interest (possibly by the use of radio waves). The time difference (or phase difference) between receivers is used to determine the location of the object. An early exponent of this was LORAN, which was a hyperbolical terrestrial radio positioning system. A terrestrial navigation or positioning system refers to one that is on the Earth. In the case of satellite-based navigation, however, an individual receiver is processing differences in the reception from multiple satellites.

WIRELESS POSITIONING SYSTEMS

The terms localization, positioning, and position location all mean the same.

A wireless positioning system is comprised of a number of nodes whose locations are known (known nodes) and one or more nodes whose positions are to be determined (unknown nodes). A node is a communication endpoint. There are GPS and local positioning systems. As its name suggests, a local positioning system does not provide global coverage. The known nodes may be at fixed locations, in which case they are known as anchor nodes. An example is beacons located on Earth. The known nodes might not be at fixed locations, as is the case with satellites. An unknown node may be stationary or it might be mobile. A base station serves as a hub of a wireless network. Fig. 1 shows a system comprised of three known nodes and one unknown node.

Types of Systems

There are different types of wireless network. We focus on those positioning and navigation systems that have resulted from extensive research and are in widespread use by individuals.

The main wireless technologies which focus on positioning are:
1. Radio-frequency identification (RFID): Used to identify or track tags attached to objects. The range of an RFID reader differs from system

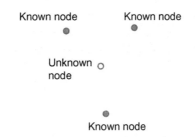

Fig. 1 A wireless positioning system.

to system. RFID is for local or regional use. At one extreme, the tag must be within 10 cm of the reader. At the other extreme, the tag can be up 200 m from the reader. In American football, some players wear an RFID tag and a player's position can be estimated to within 6 inches.

2. Radiodetermination-satellite service (RDSS): Is a service for finding the position, velocity, or similar property, of an object by using radio waves transmitted from one or more satellites. A satellite navigation system comprises a constellation of navigation satellites circling the globe on specific orbits. They send out a continuous stream of radio signals. A user of the system possesses a receiver whose function is to calculate its own position. The radio waves are broadcast at specific frequencies; any GPS radio receiver (GPSR) can pick them up. A satellite navigation system is for regional or global use. Some systems cover a region the size of a country, while others cover the globe. The principle of satellite positioning relies on accurately knowing the location of each satellite (from what is called ephemeris data) and accurate clocks. The accuracy of a satellite-based system differs from system to system. It depends on the quality of the GPSR and also on the quality of any correction that a GPSR may receive from an external source (a ground station or another satellite navigation system). With a good quality GPSR, GPS provides better than 3.5 m horizontal accuracy. (Horizontal accuracy is with reference to latitude and longitude, but not altitude.) Better accuracy can be had when a GPSR makes use of information from a fixed ground station. This is referred to as Differential GPS. It is also possible to augment GPS with another system. When this happens, centimeter-level accuracy of a GPSR's position can be achieved in real-time, and millimeter-level accuracy can be achieved when the measurement data is not processed in real-time.

In addition to these, there are other wireless technologies in which positioning is not the main use but only a by-product. The main ones are:

1. Cellular networks: A cellular network is a communication network where the last link, that is, to the end-user, is wireless. In a cellular network, a node comprised of equipment used directly by an end-user for communication is called a mobile station (MS) or user equipment (UE). Such a network is for regional use.

2. Wireless sensor networks (WSNs): Another type of wireless network is a WSN. This is used for the collection of data from sensors. In a WSN, a node that incorporates a sensor is called a sensor node. A WSN is for local or regional use.

3. Wireless local area networks (WLANs). A WLAN is for local or regional use.

Terrestrial positioning systems are dependent on a number of ground-based reference stations. Past systems that have been developed include Decca and LORAN, for the navigation of aircraft and ships. Current systems include DME, TACAN, and VOR, all for the navigation of aircraft.

Many different modulation formats and terrestrial wireless systems have been proposed for the positioning of personal handset and portable devices. These have resulted from studies of WLANs, wireless metropolitan area networks (WMANs), and WSNs. There are a variety of technologies used in RTLS, such as radio frequency identification (RFID) and ultra-wideband (UWB). Table 1 summarizes some of the available satellite and terrestrial systems. A Cell ID is a number that identifies the approximate location of the user's device relative to the cells of the network. RSSI refers to the receiver signal strength indicator which measures the received radio signal's strength. TOA refers to the time of arrival. AOA refers to the angle of arrival.

Multipath can cause errors and affect the quality of communications. A signal may be received directly but there may be others that have been reflected off adjacent features such as chain link steel fences, tall buildings, etc. Nonline-of-sight (NLOS) or near-line-of-sight is radio transmission across a path that is partially obstructed, usually by a physical object. If the line-of-sight (LOS) between two nodes is partially obstructed the main signal component is attenuated appreciable, so much so that it may be weaker than the multipath signals. In a localization system, if the LOS is fully obstructed then the received signals do not propagate through a LOS path. The main signal component can only reach the receiver after it has been reflected. This can result in inaccurate positioning data. In this case,

Table 1 Some positioning systems

Technology	Measurement technique	Notes
GNSS (e.g., Galileo, GPS)	TDOA	Accuracy: Depends which of the two services are used. For public use of GPS, high-quality receivers provide horizontal accuracy of better than 3.5 m most of the time Pros: Also provides the time. Used in all weathers. Used anywhere on or near Earth Cons: Needs unobstructed line of sight of four or more satellites
A-GNSS	TDOA	Accuracy: Same as a GNSS Pros: As for pros of GNSS. Can be used where there are weak signals Cons: As for pros of GNSS
Cellular	Cell ID/E-OTD/OTDOA/ U-TDOA	Accuracy: With U-TDOA it is 50 m
Inertial Navigation System (INS)	Angular velocity; linear acceleration	Accuracy: Depends on the type of system. Two-dimensional accuracy varies from 9 to 2200 m
NFER	Near-field properties of radio waves	Accuracy: 30 cm
RFID	TDOA	Accuracy: 10 cm Pros: No need for line of sight
WLAN	AOA/RSSI/TOA; radio fingerprinting	Pros: Used where GNSS is inadequate because of multipath, or indoor location, etc.
WSN	AOA/RSSI/TDOA/TOA	

the delay of the signal does not represent the true time of arrival (TOA), as it includes a positive bias, called the NLOS error.

An INS is quite a different positioning system to the other technologies. It makes use of items of equipment called inertial measurement units (IMUs). An IMU comprises accelerometers and gyroscopes. An IMU can be used together with GPS (augmented with the RTK system); there are several benefits to doing this.

A satellite navigation system can be augmented with another system in order to improve performance. For example, it could be augmented with a terrestrial system. Assisted GPS is a system that comprises GPS augmented with a cellular network, where usage is made of the data in the cell towers.

Hopping

Some wireless positioning systems make use of a network of nodes. Consider a set of nodes on the Earth's surface. One of the nodes is an unknown node, that is, its position is not known. The other nodes are anchor nodes, their positions are known. If the unknown node can communicate with the anchor nodes then it may be able to estimate its position. If the unknown node communicates directly with an anchor node then we say that single-hop communication is taking place. For those anchor nodes that the unknown node cannot communicate with directly, indirect communication may be available if the anchor nodes work cooperatively. In this case, multihop communication is taking place.

POSITIONING TECHNIQUES

The conventional approach to estimating the location of an unknown node involves two steps. The first step takes measurements derived from the characteristics of received radio waves. One type of measurement is the difference in the power of the received signal strength indicator (RSSI) as compared to the original signal strength. Another type of measurement is the TOA of the signal. The angle of arrival (AOA) of a signal at an unknown node is yet another type of measurement. The second step involves performing calculations on the measurements to estimate the unknown node's position. Another approach is to estimate the location of an unknown node directly.

Some position estimation approaches use a set of measurements all of one kind—AOA, RSSI, or TOA. Alternatively, AOA information may be combined with distance estimates, found using RSSI or TOA, to establish the location of an unknown node.

Measurement Techniques

Different measurement techniques can be used, such as AOA, RSSI, and TOA.

Time-of-Arrival Measurements

When time measurements can be made accurately, then time-based ranging permits highly accurate positioning. The challenges with this form of ranging are in ensuring that the transmitted signals are synchronized and in analyzing measurement errors.

Consider a situation in which node A emits a packet, containing a time stamp, and node B receives it. If the nodes are both synchronized to a reference clock, then the distance between A and B can be calculated. If there is a synchronization error, then this significantly affects the accuracy of the estimated distance.

Consider a situation in which node A emits a packet and node B receives it, as above. After an agreed upon delay, node B transmits the packet back to node A. This scheme overcomes any clock synchronization error. However, if node B's clock is drifting with respect to node A's clock, then the delay in B transmitting will be incorrect. Even a small clock drift can result in a large error in the distance estimate.

Angle-of-Arrival Measurements

This involves taking bearings. With perfect measurements, the desired location is at the intersection of the lines of bearing. In reality, measurements are not perfect due to such things as the limited resolution to which an angle can be measured, multipath, and noise. Therefore, more lines of bearing are required than with the ideal case. (In 2D, ideally only two lines would be required.)

Connectivity

Consider a network in which each node is connected to some nodes and not connected to others. We could find out which anchors are connected to the unknown node. Assuming that these anchors are closer to the unknown node than the other anchors then we could estimate the unknown node's location.

When positioning, we usually need to know which geodetic datum we are using, that is, the coordinate system used in the vicinity of the Earth. However, in certain situations, such as a military maneuver or a search and rescue operation, it may be sufficient to use relative coordinates.

Positioning in Cellular Networks

There are a number of measurement techniques used in cellular networks. Enhanced observed time difference (E-OTD), OTDOA, and U-TDOA are based on time difference of arrival (TDOA). E-OTD was first implemented on mobile phones adhering to the GSM standard. The techniques are similar, however, with the E-OTD and OTDOA measurements are made by the handset whereas with U-TDOA the measurements are made by the network. With E-OTD, a handset observes the time difference of signals sent from two base stations. OTDOA is used with UMTS, which is a system that adheres to the GSM standard. A handset observes the time difference of signals sent from several enodeBs (components of the network). With U-TDOA, a signal is sent from the handset and is picked up by location measurement units (LMUs). A handset location is based on the time taken for the signal to reach each LMU.

Positioning in Wireless Sensor Networks

An example of a measurement is the distance between an anchor node and a sensor node. This process of determining the distance from one location to another is called ranging. Wireless positioning systems can be classified according to the network configuration and the type of measurement that is carried out.

Lateration

One approach is to use a number of fixed anchor nodes at known positions. The anchors are synchronized to emit a signal at the same time. Let us consider a WSN. The sensor node is at an unknown position. It picks up signals from the anchors and works out its position using the TDOA technique. (The technique is similar to the way in which a GPS receiver calculates its position.) Fig. 2 illustrates the approach. Another approach is for the sensor node to broadcast the signal. The signal is received by each anchor, which calculates its TOA. The TOAs are passed to the base station and it calculates the sensor node's position. Fig. 3 illustrates the approach. The base station can be connected to the anchors using wires. Consider the first approach, where the sensor node receives three signals. Let the three anchors be A, B, and C. Let the unknown times taken for the signals from A, B, and C to reach the UE be t_1, t_2, and t_3, respectively. We know the difference is time between the arrival of the signal sent from A and the arrival of the signal sent from B. Call this $\tau_{1,2}$. The equation $|t_1 - t_2| = \tau_{1,2}$ describes a hyperbola. Similarly, the equation $|t_1 - t_3| = \tau_{1,3}$ describes a hyperbola. The intersection of these hyperbolae marks two positions, one of which is the location of the sensor node, as shown in Fig. 4. If the coordinates of

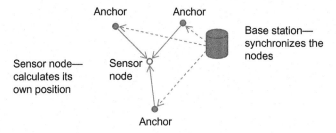

Fig. 2 TDOA approach 1.

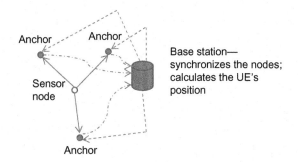

Fig. 3 TDOA approach 2.

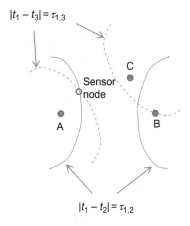

Fig. 4 Lateration by TDOA.

the anchors and sensor nodes are given in 3D then there needs to be four anchors in order to determine the location of the sensor node.

AOA Measurements

In one approach, an antenna on the sensor node estimates the AOAs of signals sent from the anchors. In another approach, an antenna on each anchor estimates the AOA of a signal sent from the sensor node.

Connectivity

One approach for positioning is where the sensor node calculates its position using signals transmitted from anchors. Another approach is where the sensor node transmits a signal and the network of anchor nodes works out the sensor node's position and sends the result to the sensor node. Calculation of the position of a node in a large network is more difficult that it is in a small network, as the complexity of the computations increases with network size. A node's position can be calculated by a central unit and done using a distributed method. Distributed methods are more scalable as the task can be distributed over the whole network. The position of a node can be calculated using one of the mathematical methods: multilateration, trilateration, and triangulation. The data used in these methods is one of the properties of the RSSI, TOA, TDOA, and AOA.

The IEEE 802.15.4 Physical Layer

IEEE 802.15.4 is a standard for low-power, low data rate wireless communication between small devices. 802.15.4 is a PHY and MAC layer protocol (OSI layers 1 and 2). Devices are segregated into personal area networks (PAN). Each PAN has a PAN Identifier—16-bit number. Each device has two addresses: a long address (64-bit globally unique device ID) and a short address (16-bit PAN-specific address). Device addresses are used in positioning. Networks can be built as either peer-to-peer (P2P) or star networks. The topology impacts upon how positioning is carried out. Several PHYs are available. One of these uses UWB. A recent application area is precision locating and tracking. UWB is used for real-time location systems; its low power and precision make it well suited for environments that are radio-frequency-sensitive environments, such as hospitals. The positioning technique to be used could, for example, be fingerprinting. A possible method is to use one based on the maximum likelihood (ML) principle. Another technique, one that estimates the TOA, is based on a frequency domain estimation algorithm.

Interferometry

Many radio interferometry (RI) positioning systems require no special hardware. The average error in many systems is less than a meter. The principle on which RI positioning works is that two independent transmitters transmit sine waves at slightly different frequencies, and this causes a radio interferometric signal to be generated. This composite signal is received by a third node. The envelope of the received signal has a beat frequency equal to the frequency difference of the sine waves. The envelope can be precisely controlled (in the short term) and so it can be adjusted to take account of the available time synchronization accuracy, limited processing capability, and sampling rate of the wireless sensor nodes.

Positioning Approaches for WSNs

There are a variety of position approaches for WSNs, some of which are ML estimator, least squares (LS) approximation (both nonlinear and linear), projections onto convex sets, multihop localization, range free localization, and semidefinite programming (SDP). LS approximation is commonly used by researchers as a benchmark with which to compare algorithms. If measurement errors are independent and identically distributed (i.i.d.) Gaussian then the position calculated by the weighted nonlinear least-squares approach is the same as the position calculated by the ML approach.

Range-Free Positioning

Range-based positioning uses distance or angle measurements to determine position. Range-free does not use distance or angle measurements. One approach is for GPS-enabled nodes to send messages to the node whose position is to be estimated.

Semidefinite Programming Algorithms

One positioning approach is named the Constrained semidefinite programming localization algorithm (CSDPLA) and differs from other approaches to the problem. Instead of minimizing the error in the estimate of the distance between a pair of nodes, it uses a technique called SDP, which involves using information about connectivity. The idea is to reduce the dependence of a node's position error on the errors in distance estimates. Some positioning approaches have the nodes cooperating with one another.

Positioning in Wireless Local Area Networks

The existing positioning techniques are AOA based, fingerprinting based, RSSI and lateration based, and time of flight (ToF) based. Radio fingerprinting can be used to identify a device by the fingerprint that characterizes its signal transmission. A WiFi positioning system is used where GLONASS or GPS are inappropriate for reasons such as multipath and indoor usage. Accurate indoor positioning is becoming more important due to the increased use of augmented reality, health care monitoring, inventory control, personal tracking, social networking, and other indoor location-aware applications. The most common technique used for positioning with wireless access points is based on fingerprinting. This uses RSSI. However, if the application is the tagging and locating assets in buildings, that is, room-level accuracy (approx. 3 m), the ToF is suitable. Fingerprinting makes use of a database in which the coordinates of known positions are stored. For each position, the signal strength to each of several nearby access points is also stored.

Time of Flight Based

The user device receives a signal from each wireless access point, whose location is known. From the timestamp of a signal the user device can calculate the ToF of a signal and use this information to estimate the distance to the access point that transmitted the signal. After receipt of a signal from each of a number of access points, the relative position of the user device with respect to access points can be established. The error in the time measurements is in the order of nanoseconds and the resulting positioning errors have been found to be in the order of 2 m.

The Ultra-Wide Band (UWB) Technology

Recent applications of UWB include precision locating, sensor data collection, and tracking. Due to low emission levels, UWB systems are used in short-range indoor applications. A Federal Communications Commission (FCC) report and order, on February 14, 2002, authorized the unlicensed use of UWB in the frequency range 3.1–10.6 GHz. More than four dozen devices have been certified under the FCC UWB rules, most of which are for radar, positioning, and imaging applications.

The General Positioning Problem

Measurements are taken from which a position is estimated. If the measurements are exact then the problem can be solved deterministically. One method is to calculate the intersection of spheres. Another is to calculate the intersection of hyperboloids. In reality, however, there are measurement errors and so alternative solution approaches are used. When there are no measurement errors, we can perform positioning by making the minimum number of measurements of some parameter(s). When there are measurement errors, we can take even more measurements to try to find the position of best fit with respect to all of the measurements. The more measurements taken, the more accurate will be the position estimate. The general positioning problem is now described. There are a number of wireless nodes, one of which is located at an unknown position. Let $\mathbf{x} = (x_0, y_0, z_0)$ denote the unknown node's position. We want to find an estimate of the unknown node's position, $\mathbf{x'}$, based on a set of measurements, $\mathbf{r}$. The measurements are taken with reference to known nodes. Example measurement techniques are AOA, RSSI, and TOA. The objective is to calculate such that a metric is minimized. One such metric is root-mean–square deviation (RMSD). Fig. 5 shows a schematic of the above problem description.

Consider a problem in two dimensions. There are three known nodes. There is one node whose position is to be found. The coordinates of known node i is (x_i, y_i). The true distance from node i to the unknown node is d_i. The unknown node's position is (x_0, y_0). The following system of equations holds:

$$(x_1 - x_0)^2 + (y_1 - y_0)^2 = d_1^2$$
$$(x_2 - x_0)^2 + (y_2 - y_0)^2 = d_2^2$$
$$(x_3 - x_0)^2 + (y_3 - y_0)^2 = d_3^2$$

Each equation represents a circle. The system of equations can be solved to find (x_0, y_0). In reality, there are measurement errors; the measured

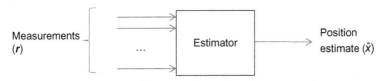

Fig. 5 The positioning problem.

distance from node i to the unknown node is δ_i. The estimate of unknown node's position is (x_0', y_0'). In order to find this estimate we use more than three equations, N_{Knodes} equations. We can express this new system of equations as:

$$\mathbf{A} \cdot \mathbf{p} = \mathbf{b}$$

where

$$\mathbf{p} = \begin{pmatrix} x_0' \\ y_0' \end{pmatrix}$$

We can solve this system of equation using the linear LS approach. This gives:

$$\mathbf{p} = (\mathbf{A}^T \mathbf{A})^{-1} \mathbf{A}^T \mathbf{b}$$

To get the system of equations into the form $\mathbf{A} \cdot \mathbf{p} = \mathbf{b}$, we can subtract last equation of a circle from every other equation, and then rearrange the equations into a system N_{Knodes} equations. Note that if the measured distance from the last node to the unknown, δ_{last}, has a large error then the error will be propagated throughout the calculations.

Let us return now to the general three-dimensional case. If $\mathbf{x}$ is the true position of an object, and $\mathbf{x}'$ is the estimated position, then the error in the estimate is given by:

$$e(\mathbf{x}) = ||\mathbf{x}' - \mathbf{x}||$$

When several estimates of a position have been made, a measure of the difference between the estimates and the true position is given by the RMSD:

$$\text{RMSD} = \sqrt{\frac{1}{n} \Sigma_{i=1}^n e_i^2(x)}$$

RMSD is often used as a measure of accuracy.

Let $\mathbf{r}$ be a set of measurements. The minimum mean square error estimator of $\mathbf{x}'$ is:

$$\mathbf{x}'(\mathbf{r}) = \int \mathbf{x} \cdot p(\mathbf{x}|\mathbf{r}) \, d\mathbf{x}$$

Statistical Techniques

Parametric

These techniques assume that the sample data comes from a population that follows a probability distribution based on a fixed set of parameters.

Nonparametric

These techniques do not assume the set of parameters to be fixed. An example technique is the LS method. The expression that is the sum of squares, which we are trying to minimize, is called the LS estimator.

LIMITS IN THE ACCURACY OF WIRELESS POSITIONING

The advent of GPS-enabled cellular phones, and advances in WiFi positioning, have meant that we have location-awareness. However, there are problems in environments where there is multipath and also in indoor residential environments. Applications for these harsh environments require submeter accuracy. We are striving for ubiquitous high-definition situation-aware (HDSA) applications. Ultra-wide bandwidth signals may contribute to this ubiquity. Cognitive radio (CR) may also contribute. This is a wireless transceiver that can detect which communication channels are in use and which are not, and communicate via a vacant channel.

Lower Bound in Parameter Estimation and Positioning

In statistics, an estimator is a rule for calculating an estimate of a given quantity based on observed data. The quantity of interest is called the estimand and its result is called the estimate. One of the properties of an estimator is whether or not it is biased. The Cramér-Rao bound (CRB) expresses a lower bound on the variance of estimators of a deterministic parameter. In its simplest form, the bound relates to unbiased estimators. An unbiased estimator which achieves this lower bound has the lowest possible mean squared error among all unbiased methods.

Some Limits for UWB Signals

An understanding of the theoretical performance limits of TOA estimation is important in the design of TOA estimators. In an AWGN channel (in the absence of other error sources), the received signal can be written as:

$$r(t) = \sqrt{E_p}p(t - \tau) + n(t)$$

where E_p is the average received energy, τ is the TOA, $p(t)$ is a unitary energy pulse transmitted through a channel, and $n(t)$ is additive white Gaussian noise (AWGN) with zero mean and two-sided power spectral density $N_0/2$.

We are interested in estimating the TOA of the direct path, τ, by observing the received signal $r(t)$ within the observation interval $[0, T_{ob})$. The problem of TOA estimation is a classical nonlinear parameter estimation problem. The solution is based on a matched filter (MF) receiver.

Let us consider the Cramer-Rao lower bound (CRB). For UWB systems we consider the nth derivative of the basic Gaussian pulse $p_0(t) = exp(-2\pi t^2/\tau_p^2)$. We have

$$p(t) = p_0^n(t)\sqrt{\frac{(n-1)!}{(2n-1)!\pi^n\tau_p^{(1-2n)}}}$$

for $n > 0$ where τ_p is a parameter affecting the pulse width and $p_0^{(n)}(t)$ denotes the nth order derivative of $p_0(t)$ with respect to t.

Innovative Positioning Techniques

The direct position determination method is an alternative to the traditional two-step location methods. It omits the process of estimating the measurement parameters. Another approach involves P2P networks. Nodes can communicate with their neighbors. By exchanging data, each node helps its neighbors to compute their positions. This approach is termed cooperative positioning. Finally, CR needs an algorithm to identify the best available wireless channel.

Direct Position Estimation in GNSS

A GNSS receiver computes its position by a two-step procedure. First, parameters are estimated and, then the estimates are used to obtain the receiver's position. Direct position estimation (DPE) is a possible alternative. It involves computing a receiver's position directly from the GNSS signal. This is a single-step procedure. It may be possible to mitigate the problem of multipath using DPE.

An example of DPE work that has been done is where the location of a transmitter is to be found by moving receivers. With one approach, the transmitter's position is calculated based on only the Doppler effects. With another approach, usage is made of the Doppler effects and the relative delays of the transmitted signal. This DPE work is unsuited to GNSS, as the problem is to find the location of a receiver.

INNOVATIVE POSITIONING TECHNIQUES

There are now MIMO WiFi interfaces, comprising of multiple antennas. A receiver receives signals via multiple paths. It is possible to estimate the

AoA of each of these signals and calculate the location of the receiver by triangulation.

One of a CR's features is its ability to determine its location. The spectrum that has been assigned is under-utilized; CR goes some way towards improving utilization. CR can be used over certain frequencies provided it does not cause noticeable interference to the assigned service.

Algorithms Based on Projections Onto Convex Sets (POCS)

Assume we wish to find the position of a node in a WSN. The node transmits a signal which is receiver by several nodes at known locations. We would need to use a property of the signal, for example, received signal strength (RSS). A weighted least squares (WLS) estimation of the unknown position can be found. Implementation of such an algorithm may be highly complex, and the presence of local minima and saddle points in the WLS objective function can cause problems. The algorithm could converge to a local minimum.

To overcome the problem of convergence to a local minimum, another type of algorithm can be used, one based on projections onto convex sets. After a few iterations such an algorithm converges to a global minimum.

Normally, algorithms based on RSS measurements need to know, or be able to estimate, channel parameters (such as path-loss exponent and transmission power). However, these parameters may be unknown and difficult to estimate for each anchor node.

Cooperative Positioning

In P2P networks, nodes can communicate with their neighbors without central coordination. There are various estimation algorithms such as LS, Kalman filter (KF), particle filter, and sum product.

Least Squares

Consider a wireless network with N nodes. Let $\mathbf{x}_i$ be the position of node i in the network and $\mathbf{x}'_i$ its estimated position. Let $S_{\rightarrow i}$ be the set of nodes that node i can receive signals from. Consider a node $j \in S_{\rightarrow i}$. An estimate of the distance between nodes i and j is $d'_{j \rightarrow i} = ||x_i - x_j|| + n_{j \rightarrow i}$, where $n_{j \rightarrow i}$ is the measurement noise.

Seamless Positioning

A seamless positioning system is one that uses multiple technologies seamlessly. For example, GPS works outside but not inside buildings or underground. If we enhance a receiver so that it uses GPS outdoors and another technology indoors then we have a seamless positioning system.

Coping With NLOS

When there is NLOS, the TOA of the first detectable signal at the unknown node is longer than it would be had there been direct LOS, resulting in a positively biased range measurement. KF preprocessing can be applied on the received TOA data for NLOS mitigation. Furthermore, NLOS mitigation can be done by using an extended Kalman filter (EKF). In addition, the state vector of the EKF, comprising of positions and velocities, can be amended by adding NLOS biases. This further helps in improving the positioning accuracy when there is NLOS.

FURTHER READING

Dardari, D., Luise, M., & Falletti, E. (Eds.), (2011). *Satellite and terrestrial radio positioning techniques: A signal processing perspective*. Amsterdam: Elsevier.

REFERENCE

St. John's College, University of Cambridge. (2014). *The way to the stars: Build your own astrolabe*. Retrieved from http://www.joh.cam.ac.uk/library/library_exhibitions/schoolresources/astrolabe/build/. (Accessed 26.04.16).

CHAPTER 2

Introduction to GPS

SATELLITE-BASED SYSTEMS

We now describe the difference in technology between a satellite positioning system and a terrestrial positioning system.

Mathematics is important to the deployment and usage of satellites. There is a combination of different types of mathematics. Processing digital data extracted from satellite signals requires deterministic mathematics. These signals pass through the atmosphere and fluid mechanics is used to model that. As many of the calculations are approximations, the mathematics of uncertainty comes into play. As an example of the relevance of mathematics, take Explorer 1 which was the United State's first satellite, of any sort. It was launched in 1958. It was a rotating satellite, with rotation taking place about a certain axis. Unfortunately, the rotation became unstable and the satellite ended up rotating about the wrong axis. At the time, the mathematics used to analyze the intended rotation, the 200-year-old Euler's theory, had not predicted the instability. Nevertheless, the satellite was still able to complete its mission, and it discovered the Van Allen radiation belts. However, some satellites are critically dependent on their orientation and thus, if instability occurs, the mission fails. The problem encountered with Explorer 1, was the impetus to find an improved way of assessing the stability of a rotating satellite. These efforts have recently resulted in an extension to Euler's theory.

Uncertainties in Global Positioning System's (GPS) positioning: A mathematical discourse concerns itself mainly with GPS as it is the dominant global positioning and navigation satellite-based system. It is known for its versatility. Much of what is written is relevant to other satellite-based positioning systems.

Today, we all know what a GPS receiver is: it communicates with a satellite system and lets you know where you are on a map. A receiver receives signals from several orbiting satellites and processes them. The receiver has a built-in map. A GPS receiver calculates its position on

Uncertainties in GPS Positioning
http://dx.doi.org/10.1016/B978-0-12-809594-2.00002-2

Earth in three-dimensional coordinates, which can be converted to latitude, longitude, and altitude. A receiver can be used anywhere on Earth, at any time of the day or night, and in any weather conditions. A receiver needs to have line-of-sight with a satellite for it to receive its signal. This means that the route from the receiver to the satellite cannot be obstructed, so it cannot be used under a bridge, for example. Radio waves cannot pass through bridges, buildings, soil, trees, walls, water, and so on.

A GPS receiver used in a vehicle is not suited to some environments, such as indoor parking areas and tunnels. In these environments the receiver does not have lines of sight with the satellites. In street canyons, a receiver may not have lines of sight with the satellites and even if it does the signals could be corrupted due to multipath. To overcome the above problems, it has been suggested that wireless local area networks could be used (WLANs).

A receiver is a passive device, it simply receives signals. An unlimited number of receivers can process GPS signals at the same time without fear of overloading the system. This is analogous to broadcast TV and radio, where millions of people can receive signals without any degradation of the broadcast.

The scientific and technological aspects of satellite-based positioning systems are highly complex; however, a user can make use of a system without knowledge of these aspects.

Prior to GPS, precise positioning was often accomplished using either inertial guidance systems or low-altitude satellites. Global Navigation Satellite Systems (GNSSs) are satellite positioning systems (sat-navs). The acronym GNSSs is in common usage and refers to all past, existing, and planned systems. The first navigation satellites were developed in the late 1950s, and were used to navigate aircraft and ships. In 1958 the US Navy created a satellite navigation system called TRANSIT whose purpose was to update the inertial navigation systems used by nuclear submarines. Other early systems include, SECOR, SIKADA, and TIMATION. Several navigation satellite systems are under development or operational. Some are competitors to GPS and some could augment GPS. Some of the satellite constellations that are currently in use for satellite positioning or under development are:

1. COMPASS of China (also known as Beidou in Chinese). China operates the BeiDou-1 system for regional use. It is an experimental navigation system. Unlike Galileo, GLObal NAvigation Satellite System

(GLONASS), and GPS, BeiDou-1 uses satellites in geostationary orbit. China plans to extend BeiDou-1 to become a GNSS (Beidou 2).

2. GALILEO of the European Union (EU), operated by the European Space Agency. Galileo is a GNSS that is under development.

3. GLONASS of Russia, created by the former Soviet Union and now operated by the Russian Aerospace Agency. It is similar to GPS in its architecture. GLONASS is an operational GNSS that is in the process of being modernized.

4. GPS of the United States. GPS is operational and in the process of being modernized.

5. IRNSS of India. India is developing the system for regional use.

6. QZSS of Japan. Japan is also developing a system. It complements GPS. It will be for regional usage.

GPS was the original GNSS system and is the dominant system. The impetus for developing new GNSSs is due to GPS's success when used for positioning.

Receivers are available that determine position using signals from more than one GNSS.

GPS

In 1973 the Pentagon proposed a second-generation guidance, navigation, and positioning system, a global satellite navigation system. One of the reasons for the proposal was the absence of a system that a receiver could use at any time of the day. GPS was developed as part of a military satellite-based navigation system. The United States Department of Defense (DoD) wanted to use GPS as part of the NAVSTAR program for highly accurate navigation using radio-based ranging. As GPS is controlled and operated by the military, a number of its aspects are classified. The launching of satellites commenced in 1978, nearly 40 years ago. Initially, GPS was for military use but in 1983 the US President announced that it would be available for civilian use once completed. By the mid-1980s, GPS had evolved to the extent that it possessed many of its present day capabilities. GPS was partially operational by 1993 and fully operational by 1995. The Federal Radio Navigation Plan stipulated that GPS was to be the US Government's main navigation system. The year 2015 saw the 20th anniversary of it being fully operational. Today, the network of satellites is called NAVSTAR-GPS (Navigation System Using Timing and Rangin-Global Positioning System).

The general public refers to it as GPS, whereas the military refers to it as NAVSTAR. It has become an accurate and stable long-term reference. GNSSs, such as the GPS, are currently the most accurate positioning systems available to navigators. GPS was quickly adopted by civilian users for a wide variety of positioning and navigation applications. Today, many millions of devices, down to smartphones, use GPS navigation. No charge is levied for making use of the satellites' signals.

Context and Applications

Satellite navigation systems have numerous civilian uses, such as:

1. Farming.
2. Navigation:
    a. Walking—using hand-held devices. GPS is a familiar tool for backpackers. Suitable GPS-enabled devices are inexpensive and their functionality provides a huge supplement to that of a compass.
    b. Driving and other transportation uses—using devices installed in aircraft, cars, trucks, and ships (see Fig. 1). Civil aviation is rapidly adopting GPS utilization due to the advantages it possesses over other means of navigation.
    c. Emergencies—search and rescue.
3. Surveying.
4. Location-based services (LBSs).
5. Map making and the provision of data to a geographic information system (GIS).
6. Gathering sports data.
7. As a clock and to perform synchronization.
8. Geophysical sciences.

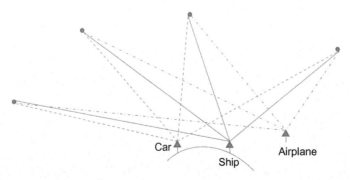

Fig. 1 Use of GPS in transportation.

9. Wildlife management.

10. Munitions—smart bombs or precision-guided munitions.

The applications evolved at a rapid rate.

As an example, farm tractors attached to a seed drill are available that allows a farmer to plant seeds accurately in a field. The farmer can position the implements to within 4 inches. As regards the user interface of the receiver in the tractor cab, the farmer could have a perspective view of the rows where he/she is intending to sow the seeds. The row in which sowing is currently taking place could be highlighted and the position of the tractor on that row shown. Around this visual display could be a variety of buttons for functions relevant to this application.

SURVEYING

Surveyors and engineers routinely use satellite surveying systems on site. GPS receivers used by surveyors are more sophisticated than the handheld ones used by the general public. Surveyors' receivers are usually pole-mounted. A position is displayed to an accuracy of anything up to a few centimeters, or better. For less-demanding surveys, a low-order rover survey receiver could be used, giving submeter accuracy. For a construction site, a high-order roving receiver could be used, giving centimeter or millimeter accuracy. A GPS antenna can be mounted on an adjustable-length pole, a bipod, or a tripod. The vertical height (antenna reference height (ARH)) is calculated using Pythagoras' theorem:

$$ARH = \sqrt{SlantHeight^2 - AntennaRadius^2}$$

A place where the coordinates are to be determined is called a new station. For each new station, all pertinent information is recorded (see Fig. 2). This includes the equipment numbers, the operator, the project number, the session times, and so on. There are different surveying techniques: traditional static, rapid static, reoccupation, kinematic surveying, and real-time kinematic (RTK) surveying.

The coordinates of relevant points can be uploaded from computer files prior to the surveyor going out to the field. When in the field, the surveyor's initial task is to position the antenna pole over the point whose coordinates are to be determined. ARH is then measured, entered into a field log, and entered into the receiver. The accuracy of position at the location is displayed on the receiver.

Project number _____

Receiver model/No   _____

Data logger type/No. _____

Antenna model/No.   _____

Operator _____

Start day/Time   _____

End day/Time   _____

Fig. 2 GPS field log.

GPS can be used in many different areas. An example area is open-pit mines, where quantities of material are to be determined. Another example area is deformation studies. This could involve geology, such as the movement of tectonic plates. It could also involve monitoring the stability of bridges and dams. Yet another example area is aerial surveying of land and water. A GPS receiver is located in an aircraft and is in contact with ground-based, or water-based, stations. When used with an Inertial Navigation System (INS), there is no need for external stations.

When changes to the land are being made, the engineer needs to know the existing height of the ground at a point and also what the proposed height is to be. GPS receivers can be standalone or can be integrated with other equipment, such as an inertial measurement unit (IMU). Portable equipment might come in a backpack or be handheld. When building structures, millimeter accuracy of vertical dimensions may be required and in this case GPS receivers can be integrated with laser devices.

Let us look at RTK in more detail. It makes use of differential positioning. It combines GPS receivers, mobile data communications, on-board applications, and on-board data processing. Its advent led to a new era in surveying. RTK makes use of a receiver at a base station and a roving receiver carried by a surveyor. Both receivers simultaneously track the same satellites. In addition, the satellite signals received at the base station are retransmitted to the roving receiver. Fig. 3 shows the arrangement.

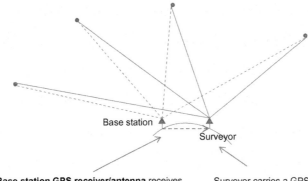

Base station

Surveyor

Base station GPS receiver/antenna receives satellite signals and hands them to a **base station radio transmitter** that broadcasts them.

Surveyor carries a GPS antenna (for receiving the satellites' signals).

Surveyor also carries a backpack containing a receiver connected to the antenna, as well as a radio receiver and radio (for receiving the base station's signals).

Fig. 3 RTK surveying.

Communication between the base station and the surveyor can be facilitated with the use of cell phones. At the start of the surveying session, it is not necessary for the surveyor to know the vector between the base station and his/her position (called the baseline). The surveyor carries an adjustable length pole on top of which is mounted an antenna and a data collector. The surveyor wears a backpack which contains a receiver, for receiving the satellites' signals, and a radio and radio antenna, for receiving signals from the base station. The RTK technique has benefitted from significant technological advances and this gained the technique acceptance among surveyors.

The data collector (a.k.a. survey controller) used in an RTK survey eases the surveyor's job. It has a number of applications, for example, cut and fill.

LOCATION-BASED SERVICES

Those who enjoy walking can use a handheld GPS receiver to find their location on a map; motorists make use of dashboard-mounted devices, of which there are a number of versions.

A main area of application of positioning is in transportation. Some of the so-called intelligent transportation systems (ITSs) make use of GPS. ITSs can be used in assisting those with broken-down vehicles, in monitoring the location of cargo, in routing, in the management of accidents, and so on. Ships use GPS receivers. Apart from transportation, handling emergencies is an important application of positioning. When using the 112 emergency telephone number, an EU directive requires mobile phone operators to provide the emergency services with details of the location of the user, if they know it. Similarly, in the United States, when a call to 911 is made, the US Federal Communications Commission requires mobile operators to relay location information to the emergency services. The commercial sector has become greatly interested in LBSs. LBSs are available on mobile devices. LBSs can be used in the open and also inside premises. Examples of LBSs include tracking of static resources, using RF tags, and tracking of people or things that are moving. Unfortunately, GNSS cannot be used for indoor positioning. This is because GNSS requires line of sight with satellites as the signals are attenuated by solid objects. Solid objects also cause multipath. An indoor positioning system can make use of one of a number of wireless technologies, such as UWB. The existence of LBSs means that marketing and advertising strategies need to be rethought so as to make them local to the consumer. Organizations can track their off-site employees, who will be using a navigation aid. The ease with which a person or vehicle can be positioned has opened up the possibility of location-specific billing. Fig. 4 shows a schematic that gives the basic interaction between a user and a context-sensitive service provider.

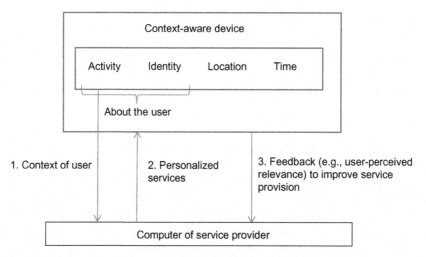

Fig. 4 User interaction with context-sensitive service provider.

In future, GPS will be usable in harsher environments, such as within buildings and in tunnels. This brings with it a whole raft of new applications such as tracking of persons (within a building for security purposes, hospital patients, and fire-fighters) and locating assets (finding an item of medical equipment in a hospital). A system which tracks a person or locates an asset in real-time is termed a real-time location system (RTLS). It is predicted that there will be tremendous growth in RTLSs.

MAP MAKING

In the modern age, people are increasingly becoming dependent on road maps; consider the popularity of Google Maps, for example. The manual upkeep of these maps is prohibitively time-consuming. Maps need to be accurate and this means that they must be response to changes to roads. Research has been undertaken to use GPS data, and inference, to automatically update maps. This data could be taken from users' GPS receivers.

To date, the research has focused on correcting the geometry of roads. However, another area of difficulty concerns road intersections. Problems include such things as roads missing, no entry roads unmarked, and roads closed. Road intersection problems occur more frequently than do geometry problems. However, the former problems are more difficult to infer than the latter. Research has been undertaken in Shanghai to automatically update road intersection information (Wang, Forman, & Wei, 2015). Over a period of 21 months, the GPS data from more than 10,000 taxis was processed.

The Shanghai research (Wang et al., 2015) involved the use of an established algorithm. Simple algorithms were then proposed for identifying errors in maps—roads missing, no entry roads unmarked, and roads closed.

SPORTS DATA

One area involves gathering big game data, that is, soccer data, and using a program to analyze this data. Players wear GPS receivers and data is collected during a match. The wearing of GPS units is not allowed by some governing authorities but may be acceptable at lower levels, all the way down to training sessions. Professional football clubs employ Performance Analysts whose job is to use such a program. An analyst's job is to get simple statistical analyses in order to help players, coaches, and soccer club owners.

Companies developing such software have existed since the late 1990s. They develop technology that exploits the potential for gathering data, and analyzing it, in order to assist those making decisions in professional soccer. An example company is Prozone. A program could provide support to a club at the various levels: first team, reserves, and those in any academy. At each level one would expect a different level of performance. Over a period of time a club collects historical data and this can be used to benchmark players. A club can then compare the statistics for an individual player with existing and previous players.

GPS data can supplement data from other sources such as heart rate monitors. More generally, an overview of a player can be assembled from his/her health monitoring data, medical history, salary history, and income above and beyond his/her salary.

In rugby, players use wearable technology—GPS receivers in the back of their shirts.

A cyclist could use a GPS receiver to gather data about himself/herself. The data is uploaded into either a mobile or online app. The purpose of the app is to help the cyclist see how well (or how badly) he/she is performing. Thanks to the Tour de France and other events, as well as the realization that bikes are a healthy form of transport, cycling appears to be gaining in popularity. Example groups of cyclists are:

1. professional cyclists,
2. middle aged men in lycra (MAMILs), etc.

A relevant app to this application is Strava. A cyclist can compare his/her performance with that of other cyclists. A cyclist can upload GPS data to confirm that he/she took a certain period of time to complete a specific route. The cyclist can then check a league table to see how he/she ranks, compared to others, for cycling that route.

Runners make use of GPS-enabled watches connected to sensors in their shoes. After a run, a runner could upload the gathered GPS data into a program that shows the run on a map, indicating inclines.

UNCERTAINTY IN GPS POSITIONING

If the position calculated by a GPS receiver is erroneous, too inaccurate, or not available then this could have serious repercussions. Fortunately, over its long history, GPS has usually proved very successful. GPS receivers are normally reliable, working correctly for many years without maintenance. There have been satellite malfunctions and cases of signal interference. Records for the deliberate interference of GPS signals are, obviously, difficult to come by. There have been many cases of unintentional interference. One case occurred in December 2001 at Moss Landing, California. A jammer was unintentionally left on causing GPS failure within a 180 nautical mile radius of Mesa, Arizona. A handheld jamming device (a.k.a. a personal privacy device (PPD)) can block GPS and mobile phone signals within a 20 meter radius. When GPS signals are unavailable, a GPS receiver goes into mode where it tries to predict the receiver's position.

Automobile drivers frequently encounter problems due to car navigational equipment. A GPS receiver in an automobile comprises two parts, a system for processing satellite signals and a map system.

GNSSs USAGE PATTERNS

User equipment is in a continual stage of development. Most receivers are either built into mobile phones or located in automobiles.

NONPOSITIONING USES OF GPS

A receiver can calculate its position, its velocity, and the true time to a high accuracy.

Some users are only interested in the clock time broadcast by the satellites. We must not forget these users. They use the time information in their specific areas of work and are often dependent on the accuracy of the broadcasted times.

Let us turn our attention to how GPS satellite signals are used in time and frequency metrology. Laboratories can measure time in units of nanoseconds or smaller. In contrast, much larger units of time are enough for everyday use. In those industries where accurate timing is essential, the unit of time used is somewhere between those used in laboratories and those used in everyday life; it is the microsecond (10^{-6} s). It is difficult for a human to comprehend how short a period of time a microsecond is and yet, as was mentioned, for a time metrologist, it is not that small. The critical-infrastructure applications the electric power grid and mobile telephone networks each need microsecond accuracy at thousands of geographically dispersed sites. They achieve this using GPS receivers that are used only as clocks, that is, GPS-enabled clocks. Currently, GPS is the only technology that can provide microsecond accuracy at thousands of geographically dispersed sites.

GPS is trusted as a time reference because the clocks on the satellites are highly accurate atomic clocks whose time is controlled by the United States Naval Observatory (USNO). The time shown on a GPS-enabled clock has small inaccuracies due to a number of factors. Even in the worst case, the time on a GPS-enabled clock differs from Coordinated Universal Time (UTC) by no more than 0.4 μs.

The advent of GPS has enabled critical infrastructure technologies. These are technologies that heavily rely on GPS-enabled clocks because microsecond accuracy is easy to achieve with GPS but difficult to achieve without it. Two such technologies are code division multiple access (CDMA) mobile phone networks and the smart grid.

Mobile phone operators such as Nextel, Sprint, US Cellular, Verizon, and others make use of a type of network called a CDMA mobile phone network. If one looks in the vicinity of mobile phone antennas, perhaps atop a street light pole, one can often see a GPS antenna. The fixed base stations in a CDMA mobile phone network receive and transmit signals. When transmitting, a specific base station's signal can be distinguished from other base station's signal due to the fact that each base station makes use of a unique time offset when constructing its signal. GPS provides the time reference to which all base stations are synchronized. As a mobile phone moves from cell to cell, the handover of the phone's signal from one base station to another is thus facilitated.

The electric power grid is subjected to considerable, and growing, demand from consumers, so much so that sometimes demand is close to what it is possible to supply. To prevent outages, the power grid can be made into a smart grid. In a smart grid, the state of the grid is monitored in real time. This is done by taking measurements, in a synchronized manner, at the power substations. GPS satellites provide this synchronization.

There are a number of possible backup strategies for GPS-enabled clocks. One is to use another GNSS; another is to resurrect eLoran, a radio navigation system that was shut down. Another possibility is to use the timing signals transmitted by networks. Using fiber optics, subnanosecond accuracy can be achieved. There are various ways, and complexities, with which timing is achieved in a network. However, all calculate the transmission delay between two clocks by sending a signal from a reference clock to a remote clock; the remote clock then returns the signal, and the reference clock notes the delay in the two-hop trip. Halving this delay is

Reference clock   Remote clock

Fig. 5 Process by which network calculates transmission delay.

an estimate of the transmission delay of a signal sent from a reference clock to a remote clock. The process is illustrated in Fig. 5. The remote clock is corrected whenever the reference clock synchronizes it. As a backup for GPS-enabled clocks used in a critical infrastructure application, the possible network solutions are:

1. Building a wide area network (WAN) based on fiber optics whose sole purpose was timing.
2. Building a WAN based on fiber optics whose timing was closely controlled.
3. Deploying large numbers of reference clocks, so that any remote clock is a short distance away from a local area network (LAN).

Yet another backup strategy is to deploy many thousands of atomic clocks. A final backup strategy will now be discussed. A satellite other than a GPS satellite could be used. The on-ground reference and remote clocks both receive the same satellite signal. Both clocks are calibrated to take account of the different signal delays. The time of receipt of the signal by the reference clock is passed via a network to a server as is the time of the receipt by the remote clock. The server calculates the difference in times and sends this correction to the remote clock. The above process can be performed continuously. The remote clock is referred to as a common-view disciplined clock (CVDC). For each CVDC only a small amount of data needs to be processed by the server. As a result the server can handle many CVDCs. CVDC systems exist in Japan and the United States. A fail-safe system could be established whereby each component of the system—the common-view signal, the data network, and the reference clock—each has a backup. The backup for the common-view signal could come from a GNSS satellite or a geocommunication satellite. Multiple data networks could be used so as to provide redundancy. Whenever one of the primary components of the CVDC system fails, the backup component is brought into action.

Time and Frequency Measurements

GPS Receivers

There are a number of different types of GPS receiver used in time and frequency metrology. When used for this purpose a receiver is referred to as a GPS timing receiver. There are innumerable time and frequency applications. Most receivers provide a 1 pps output and time-of-day information. Another type of GPS receiver also provides standard frequencies. These are called GPS disciplined oscillators (GPSDOs). They have many applications. For some of the more specialized measurements, two other types of GPS receiver are used. One is a common-view GPS receiver, the other is a carrier-phase GPS receiver. The latter type is designed for geodetic and surveying applications. The GPS antennas used with most receivers, in metrology, are small—usually <100 mm in diameter.

Measurement Techniques

As stated above, different types of receiver are used in time and frequency metrology. Consequently, different types of GPS measurements are made. The measurement techniques are one-way, single-channel common-view, multichannel common-view, and carrier-phase common-view. This list is ordered, starting from the technique with the most timing, and frequency, uncertainties to the least uncertainties. The frequency uncertainty for the one-way technique, for 1 day, is $< 2 \times 10^{-13}$.

One-Way

The one-way GPS measurement technique uses the signals output from a GPS receiver as a reference for calibration purposes. As with all receivers, prior to use, signal acquisition must be performed. Once this is completed, the output signal(s) can be input to the measurement system in use.

GPS satellites transmit signals that have been designed to be in close agreement with UTC and so the long-term accuracy of a GPS receiver is excellent. The only time when this was not the case was when the DoD deliberately added noise to GPS signals in order to reduce positioning and timing accuracy. This practice was called selective availability (SA). It ceased on May 2, 2000.

When a laboratory is making a measurement, it must know what the uncertainty of the measurement process is with respect to a reference. The uncertainty of the reference must be known with respect to a superior reference, etc., all the way back to the International System of Units (SI) reference. Traceability requires an unbroken chain of comparisons with references, all comparisons having stated uncertainties. An example of a traceability train is:

SI → UTC (NIST) → UTC (USNO) → GPS Broadcast Signals →
GPS Received Signals → Users Device Being Tested by Laborator

where UTC (NIST) is the variant of UTC maintained by the National Institute of Standards and Technology and UTC (USNO) is the variant maintained by the United States Naval Observatory.

Another example traceability train is:

SI → UTC (NIST) → GPS Broadcast Signals → GPS Received Signals
→ Users Device Under Test

The uncertainties for the comparisons involving SI, UTC (NIST), UTC (USNO), and GPS Broadcast Signals are very small, and are of little relevance to most measurements. The three uncertainties can be found from published documents. In the second train, NIST records time and frequency offsets for each GPS satellite. This enables it to calculate the uncertainty of the UTC (NIST)-GPS Broadcast Signals. The uncertainty of the comparison between GPS Broadcast Signals and GPS Received Signals is receiver dependent. In order to derive this uncertainty, we must be able to state a specification that the receiver will meet or exceed when correctly operated. The uncertainty of the GPS Received Signals—Users Device Under Test link is the uncertainty in the calibration procedure being undertaken at the laboratory.

Common-View

The common-view technique is a simple way of comparing two clocks located in different places. Fig. 6 illustrates the approach. The technique removes measurement errors that are common to both locations. The time on the two clocks are compared. Similarly, the frequencies of the two clocks are compared. The technique has its pedigree in making comparisons of international time and frequency standards. There are two types of GPS common-view measurements: single-channel and multichannel.

Single-channel common-view needs a GPS receiver that is capable of reading a tracking schedule, that is, informing it when to start taking measurements and which satellite to track. Published tracking schedules are available.

Multichannel common-view does not use a tracking schedule. A receiver simply records data from all the satellites in view. This results in the collection of more data than when single-channel common-view is used.

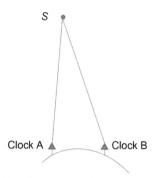

Fig. 6 Use of common-view in GPS.

A disadvantage of the common-view technique is that the data collected must be transferred between the two clocks' owners, following which it must be processed. Hence it may take some time for the measurements results to be available. To overcome this delay, some years ago NIST developed a prototype system comprised of a multichannel common-view receiver integrated with an Internet-enabled personal computer. The system could send data to a web server as soon as measurements were made.

The common-view technique works best when the distance between clocks, called the baseline, is a few thousand kilometers or less. For two clocks located on the continental United States, the time uncertainty and the frequency uncertainty, over one day and at 2σ, should be $<10\,\mathrm{ns}$ and approximately 1×10^{-13}, respectively.

As regards measurement uncertainty, the common-view technique is only slightly better than the one-way technique. However, as regards traceability, with the common-view technique, the chain between UTC (NIST) and Users Device Under Test is reduced to one link.

Carrier-Phase Common-View
This technique is mainly used for frequency measurement. A GPS satellite transmits two signals—at the L1 and L2 frequencies. Each signal is comprised of a carrier wave and superimposed on this is a pseudorandom noise code. With the techniques mentioned above, the receiver processes the code on the L1 frequency, whereas with the carrier-phase approach, the receiver processes the carrier waves on both the L1 and L2 frequencies. The carrier-phase GPS technique necessitates substantial postprocessing of the data and so is not suited to everyday measurements. International

comparisons of frequency standards, however, make use of the technique. As mentioned above, two receivers are used in the common-view approach. For international comparisons where the baseline is prohibitively long, a network of receivers is used. The data collected is processed using precise satellite orbital information and detailed models of the ionosphere and troposphere. The equation used in the carrier-phase technique is:

$$\lambda \phi_R^{S_k} = \sqrt{(x_k - x_0)^2 + (y_k - y_0)^2 + (z_k - z_0)^2} + c\tau + c\delta t_s + c\delta t_{eph}$$
$$+ c\delta t_{ion} + c\delta t_{trp} + \text{OtherErrors} + \epsilon$$

where c is the speed of light, λ is the carrier wavelength, c/f, $\phi_R^{S_k}$ is the carrier phase observable for satellite S_k and receiver R, (x_k, y_k, z_k) are the position of satellite S_k when the data was sent, (x_0, y_0, z_0) are the coordinates of the user receiver, τ is the receiver clock bias, δt_s is the satellite clock error, δt_{eph} is the ephemeris error, δt_{ion} is the ionospheric error, and δt_{trp} is the tropospheric error.

The other errors are ones such as multipath and noise. Research at NIST has shown that measurement uncertainty can be reduced by reducing the noise at both of the stations making the international comparison, by improving cycle slip detection, and by using good models of the ionosphere and troposphere.

In conclusion, GPS is the primary system for distributing highly accurate time and frequency worldwide.

Further Information

Appendix A discusses how one might calibrate a GPS receiver that is solely used for its timing pulse. A Regional Metrology Organization (RMO) called the Inter-American Metrology System, or Sistema Interamericano de Metrologia (SIM), created a comparison network to compare time and frequency standards between the member nations of SIM. This comparison network is described in Appendix B. SIM is made up of national metrology institutes in numerous countries throughout the Caribbean as well as Central, North, and South America. There are a small number of RMOs that are recognized by the Bureau International des Poids et Mesures (BIPM), one of which is SIM. Appendix C compares the characteristics of GPS receivers, as regards calibration of their delays, between several laboratories.

Furthermore, it may be possible to use a receiver as an acceleration sensor. Appendix D discusses this further.

ESTIMATING THE DISTANCE TO A GOLF FLAGSTICK

Two of the ways in which a golfer could estimate the distance to a flagstick are by using GPS and by using a laser device. Zhu & Vonderohe (n.d.) discusses the accuracies of the two approaches.

GPS

GPS does not calculate the distance between golf ball and flagstick directly. GPS can estimate the coordinates of a point on Earth by a receiver processing radio signals sent from several GPS satellites. A GPS receiver could estimate the coordinates of the golf ball and flagstick and then use them to calculate the distance between the points. The problem is that the golfer may not know the coordinates of the flagstick. The golfer could use an on-line map, possibly a satellite map, that gives the coordinates of any point selected. This will only suffice if the map is a high quality one and if the golfer is able to identify the flagstick on the map and point to it accurately. Let us analyze the errors in this approach. We would like to end up with a statement of the form "the error between the true distance and the measured distance is d meters, 95% of the time." The value of d depends on how accurate the coordinates of the ball (b) can be measured 95% of the time, and how accurate the coordinates of the flagstick (f) can be measured 95% of the time. The following relationship holds (Zhu & Vonderohe, n.d.):

$$d = 0.8\sqrt{b^2 + f^2}$$

There are a number of problems that could cause major problems for GPS positioning. These include, among others: reflection of satellite signals off hard surfaces such as the walls of a building; sunspot activity, which disturbs the upper atmosphere; weak satellite signals. Furthermore, and as was mentioned, it may be very difficult to get the desired accuracy of the flagstick position due to the quality of the map used.

Laser Ranging

A laser ranging device can measure the distance from a golf ball to a flagstick, provided that the flagstick has a prism attached to it. It would be simpler to use than a GPS receiver requiring map interaction. A laser beam spreads out as it is transmitted. The accuracy of the device is proportional to the distance being measured.

Comparison Between GPS and Laser Ranging

A laser ranging device measures the distance between a golf ball and a flagstick directly, whereas a GPS receiver does not. When comparing the two technologies, it is important to be measure accuracy in a consistent way. For example, we could use a statement of the form "the error between the true distance and the measured distance is x meters, 95% of the time" with both technologies. In conclusion, most professional golfers opt for laser ranging.

RECEIVER SPECIFICS

As there are widely different applications of GPS, there is a corresponding wide range of available receivers, differing in their functionality.

GPS receivers vary in price. The lowest precision receivers, used for day-to-day navigation, recreation (such as hiking and orienteering), and low-precision mapping/GIS are hand-held and cost from 100 to 500 dollars. Marine navigation receivers (accuracy 1–5 m) cost about 1000+ dollars. More precise receivers are used in GIS and mapping applications. With these, accuracy is improved with differential correction. For a receiver giving submeter to 5 m accuracy, the cost is between 1000 and 5000 dollars. For an accuracy of between 20 and 30 cm, the cost is between 5000 and 10,000 dollars. Receivers used in surveying have the highest accuracy (cm or mm accuracy) and cost from 5000 to 30,000 dollars.

There are many manufacturers of GPS receivers, including: Corvallis Microtechnology, Garmin International, Lowrance Electronics, Inc., Magellan Systems Corp., Motorola Solutions, Rockwell International, Trimble Navigation Ltd.

The main characteristics of a receiver are the number of channels available to it (i.e., how many satellites it can track simultaneously), the number of frequencies it can receive (i.e., either L1 or both L1 and L2), and what part of a signal it uses to calculate the distance to a satellite (i.e., either a digital code or an analog carrier wave). Some GPS receivers use GLONASS as a backup.

AN INTERESTING ASIDE

Consider three researchers who wish to share an antenna that receives signals from a satellite. Each researcher has a hut full of equipment located somewhere on a large field. The signals are transferred from the antenna to the huts by means of cables. The optimum position of the antenna, so as to minimize the length of cable used, is the Fermat-Torricelli point.

FURTHER READING

Cooksey, D. NAVSTAR Global Positioning System (GPS) facts. (n.d.). http://www.montana.edu/gps/NAVSTAR.html (Accessed 26.04.16).

National Coordination Office for Space-Based Positioning, Navigation, and Timing. GPS.GOV. (2016). http://www.gps.gov/ (Accessed 26.04.16).

Lombardi, M.A. (2012). Microsecond accuracy at multiple sites: Is it possible without GPS? *IEEE Instrumentation & Measurement Magazine*, October 2012, 14–21.

Lombardi, M.A., Nelson, L.M., Novick, A.N., & Zhang, V.S. (2001). Time and frequency measurements using the Global Positioning System. *Cal Lab The International Journal of Metrology*, July–September, 2001, 26–33.

Sharma, K.K. (2012). *Fundamentals of radar, sonar and navigation engineering (with guidance).* New Delhi: S.K. Kataria & Sons.

Uren, J. & Price, W.F. (2010) *Surveying for engineers* (5th ed.). London, New York, & Shanghai: Palgrave Macmillan.

REFERENCES

Wang, Y., Forman, G., & Wei, H. (2015). Mining large-scale GPS streams for connectivity refinement of road maps. *The Computer Journal, 58*(9), 2109–2119.

Zhu, Y., & Vonderohe, A. P. (n.d.). *Accuracies of laser rangefinders and GPS for determining distances on golf courses.* Retrieved from http://laserisbetter.com/wordpress/wp-content/uploads/2009/03/accuraciesofrangefinders-drvonderohe.pdf (Accessed 26.04.16).

CHAPTER 3

Basic GPS Principles

Astrodynamics is the field that is concerned with the properties of satellites and orbits. A satellites orbit is inclined with respect to the Earth's equator. The orbit is also either circular or elliptical (of various eccentricities). A satellite following a geostationary orbit has a 0 degrees inclination, is circular, and the speed is such that, relative to the Earth, the satellite appears stationary. A Global Positioning System (GPS) satellite, on the other hand, is in a circular inclined orbit.

Fig. 1 shows the angle of elevation of a satellite from a point on the Earth's surface. Note that as the elevation changes, so does the range. We can use trigonometry to calculate the range, d_k; see Fig. 2. Using the cosine rule:

$$(a + h)^2 = a^2 + d_k^2 2a \cdot d \cdot cos(90 \text{ degrees} + \beta)$$
$$= a^2 + d_k^2 + 2a \cdot d \cdot sin\beta$$

$$d_k = -a \cdot sin\beta + \sqrt{(a + h)^2 - a^2 cos^2\beta}$$

or

$$d_k = -a \cdot sin\beta - \sqrt{(a + h)^2 - a^2 cos^2\beta}$$

When $\beta = 90$ degrees,

$$d_k = -a + (a + h)$$

or

$$d_k = -a - (a + h)$$

$d_k = h$ is the only sensible value.

When $\beta = 0$ degrees,

$$d_k = \sqrt{(a + h)^2 - a^2}$$

Uncertainties in GPS Positioning
http://dx.doi.org/10.1016/B978-0-12-809594-2.00003-4

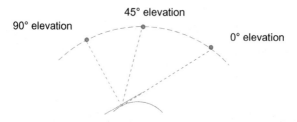

Fig. 1 The angle of elevation of a satellite from a point on the Earth's surface.

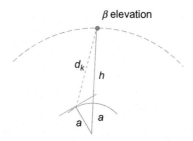

Fig. 2 Meaning of symbols used to calculate the range.

Fig. 3 shows a representative view from space of GPS orbits and typical satellite positions.

The architecture of the GPS system, like any Global Navigation Satellite Systems (GNSS), comprises three segments: Ground Segment (GS; stations positioned around the Earth to control the satellites), Space Segment

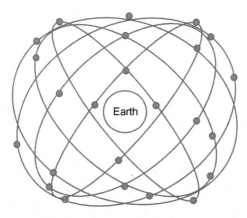

Fig. 3 GPS satellite constellation.

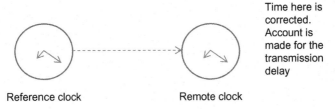

Fig. 4 Transferring time between locations.

(the satellites orbiting the Earth), and User Segment (anyone that has a GPS receiver).

CLOCKS

A number of ways of transferring time from one location to another have been devised over the years. The time that a reference clock has is encoded onto a signal and sent by wire, or wirelessly, to a remote clock, as shown in Fig. 4. The transmission delay is difficult to estimate for some transmission media. For GPS, the transmission delay can be estimated very accurately. A clock in a GPS receiver, whose time is constantly corrected by a satellite's clock, is relatively accurate, providing submicrosecond accuracy.

The accurate measurement of time and the stability of measurement are crucial to GPS. The satellites' clocks (also called oscillators) are atomic clocks, are highly accurate, stable, and are synchronized. Each satellite has up to four of them. They are either Cesium or Rubidium clocks, or a combination of both. There are chip scale atomic clocks (CSACs) that are very small in comparison to other atomic clocks. The dimensions of a CSAC are approximately $40 \times 35 \times 11$ mm and the weight is less than 35 g. A receiver's clock is an ordinary clock. This is typically a Quartz clock. There is a wide range in quality of these clocks.

A module that acts as a clock in a GPS receiver is inexpensive and can be about the size of a coin. It is thus small enough to embed in almost any electronic device.

GROUND SEGMENT

This is under the control of the United States Department of Defense (DoD). The segment is made up of a master station, ground monitoring stations, and ground antennas at various worldwide locations. The master

Fig. 5 Monitoring and/or uploading data to a satellite.

station is at Colorado Springs, Colorado. The monitoring stations are on land masses close to oceans – the Atlantic Ocean (Ascension Island and Cape Canaveral, Florida), the Indian Ocean (Diego Garcia), and the Pacific Ocean (Hawaii and Kwajalein). The monitoring stations continually monitor the satellite signal transmissions and track the satellite orbits, ensuring that they stay precisely in their specified orbits. The orbits in the short-term are predicted. There are atomic clocks in the stations. The satellites' clocks are adjusted from the stations. The time scale used is the version of Coordinated Universal Time that is maintained by the United States Naval Observatory (UTC(USNO)). GS has other functions: uploading updated navigational data (this includes the position of the relevant satellite (ephemeris)); maintaining the health and status of the satellite constellation. The current GPS satellite status and the constellation configuration are accessible via a website. Fig. 5 illustrates the communication between GS and a satellite.

SPACE SEGMENT

This is the satellite constellation. There are a minimum of 24 operational satellites (plus spares) placed in approximately circular medium Earth orbits (a.k.a. intermediate circular orbits (ICOs)). The satellites travel in six equally spaced orbital planes, four satellites per plane. The planar orbits are out of phase with one another (see Fig. 3). The semimajor axis of an orbit is approximately 26,500 km, with satellites approximately 20,000 km (about 10,900 miles) above the Earth's surface. An orbital period is 11 h 58 min. Each orbital plane has an inclination of 55 degrees. The satellites are visible overhead up to 55 degrees in both the Northern and Southern hemisphere. At least four satellites are visible from any point on the Earth's surface (with a few exceptions). Usually six to eight satellites are visible.

The number of satellites that should comprise the GPS constellation is $N = 18(R + h)/h$, where R is the radius of the Earth and h is the distance of the constellation from the surface of the Earth. Given that R is approximately 6500 km, and h is approximately 20,000 km, we have:

$$N = 18(26,500/20,000)$$
$$= 18 \times 265/200$$
$$= 23.85$$

approximately 24 GPS satellites are manufactured by Rockwell International. They span 17 feet with their solar panels deployed, and weigh about 1900 pounds.

USER SEGMENT

This segment comprises all those who use receivers. A receiver, as its name implies, is a receive-only device. It receives data from the satellites and this is used to calculate the receiver's position or velocity, or the time. Position is calculated from the estimated distances that a receiver is from each satellite (pseudodistances). A code is sent from a satellite and a replica of this code, which is stored in the receiver's memory, is generated by the receiver. The time difference between the code received from the satellite and the code generated by the receiver is used to calculate the pseudodistance. Fig. 6 illustrates this process.

Fig. 7 illustrates how the position of a receiver is calculated. A, B, C, and D are four satellites. Let d_1, d_2, d_3, and d_4 denote the true distances from

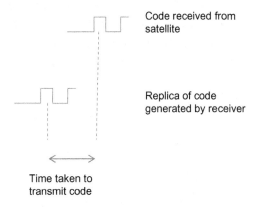

Fig. 6 Using transmission time to derive pseudodistance.

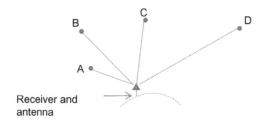

Fig. 7 Calculating the position of a GPS receiver.

the receiver to satellites A, B, C, and D, respectively. The position of the receiver is at the intersection of spheres centered at A, B, and C, with radii d_1, d_2, and d_3, respectively (or any three of the four spheres). Unfortunately, we have to make do with estimated distances and so four, or more, satellites are used to calculate the position of the receiver.

Different receivers have different capabilities and accuracies. Dual-frequency receivers are more accurate, and therefore more costly, than single frequency receivers. There are different qualities of receiver clock. Some receivers are capable of making use of the differential enhancement to GPS. Differential GPS gives users improved accuracy and is used in applications such as surveying.

Users are either civilians or from the military.

GPS SERVICES

The military uses the GPS Precise Positioning Service (PPS) while non-military users use the Standard Positioning Service (SPS). Obviously the military can also use SPS. PPS is more accurate than SPS. Both services can are single-receiver services, that is, used from a stand-alone receiver. SPS is an uninterrupted service for civilian users worldwide.

PERFORMANCE OF GPS

Targets have been set by the designers of GPS for the availability of services and the accuracy of positioning. These have been exceeded. GPS is available almost all of the time and is highly accurate.

Forssell (n.d.) compared the planned performance with actual performance, according to his 2009 article. This is now described. Some effects that influence performance are grouped together in what is called position dilution of precision. One of these is the number of satellites receivers have line of sight with. The more satellites a receiver can see, the more accurate the positioning. The design was for between four and six satellites to have line of sight from anywhere on the Earth's surface, at least 98% of the time. The actual availability of this range of satellite in early 2009 was 99.98%.

For calculating the position of a receiver (in terms of latitude and longitude, but not altitude), the planned availability was 95% of the time. When able to perform these calculations, the planned accuracy was 36 m or less, at least 99% of the time. The actual accuracy in early 2009 was 3.7 m.

For calculating the altitude of a receiver, the planned availability was 95% of the time. When able to perform these calculations, the planned accuracy was 77 m or less, at least 99% of the time. The actual accuracy in early 2009 was 5.3 m.

Forssell (n.d.) stated that the user range error, that is, the error in distance from a receiver to a satellite, as calculated by the receiver, was required to be 6 m or less. However, for the constellation as a whole, the average error in early 2009 was 1.2 m.

GLONASS

The orbital radius of satellites is approximately 25,500 km and the orbital period is 11 h 15 min. (This compares with approximately 26,500 km and 11 h 58 min, respectively, for GPS satellites.) Each of the three orbital planes has an inclination of 64 degrees (compared to 55 degrees for GPS). As a result of this inclination, GLONASS has better coverage at higher latitudes. The constellation comprises 24 satellites.

GALILEO

Prior to Galileo, EUROCONTROL, the European Commission, and the European Space Agency developed the European Geostationary Navigation Overlay System (EGNOS). Galileo is designed to be 100% interoperable with GPS and GLONASS. On May 26, 2003, the European Space Agency and the EU agreed to collaborate in order to develop Galileo. Galileo's satellites are in circular orbits 23,617 km above the Earth.

Galileo is run by civil authorities, whereas GPS and GLONASS are run by military authorities. The pre-Galileo GIOVE (Galileo In-Orbit Validation Element) satellites were aimed at testing the Galileo positioning system technologies in orbit. They comprised GIOVE-A, GIOVE-B, and GIOVE-A2. The first satellite launch in the program occurred in 2005, in which the satellite GIOVE-A was deployed. In July 2007 the United States and the EU agreed to jointly adopt a new signal design. This uses a modulation method called multiplexed binary offset carrier (MBOC). In 2008 the satellite GIOVE-B was launched, which transmitted signals created using MBOC modulation. Having proved that the key technologies are sound, Galileo satellites were launched. These commenced in 2011. The build-up of Galileo is proceeding with further satellites scheduled to be placed in orbit and further ground stations to be deployed.

Galileo's open service (OS) provides free positioning and time information. The commercial service (CS) is for professional or commercial users and offers improved performance and data with added value, as compared to OS.

REFERENCING A POSITION

The mutually orthogonal set of axes used in calculations is also known as a reference frame.

Inertial Reference Frame

As regards the axes used for the coordinate system to be used in positioning one could adopt a system where the origin is at the center of the Earth (i.e., geocentric) and the axes are fixed relative to distant stars. This means that the axes do not rotate as the Earth rotates. This coordinate system is inconvenient for use in positioning.

Earth-Centered, Earth Fixed (ECEF)

In practice the coordinate system used is geocentric but has fixed axes with respect to the Earth, and the axes rotate with Earth. It is a rotating frame of reference. The coordinates of any point on the Earth's surface are fixed. This coordinate system is termed the ECEF frame. From a standards perspective, it is called the International Terrestrial Reference Framework (ITRF). The standard coordinate system used for the GPS is referred to as WGS 84. The origin $(0, 0, 0)$ is the center of mass of the Earth, and the x-axis is a line drawn from the origin through the equatorial plane to the Greenwich meridian. The unit of time is the SI second, as specified by the equipment at the US Naval Observatory, and the unit of length is the SI meter.

Note that the vast majority of GPS users are either stationary or moving slowly on the Earth's surface. The transmitted navigation messages, and the transmission time, allow the receiver to calculate its position in the ECEF frame. Within a satellite the digital data (the navigation message) is combined with one or more pseudorandom noise (PRNs) bit sequences, and the result is converted to an analog signal. Currently, a satellite transmits two signals. PRN serves two purposes. Firstly, each satellite is given two PRNs that are unique to that satellite. A receiver can, therefore, know which signal is being broadcast from which satellite. Secondly, while a satellite is generating a PRN, the receiver is also generating the same PRN, that is, they are synchronized. Upon receipt of a PRN contained in the signal, the receiver can determine how out of sync it is with the receivers PRN, and this enables the distance between the satellite and the receiver to be calculated.

Geographic Coordinates

For simplicity, we could assume the Earth to be a perfect sphere. We could also make use of spherical coordinates. The usage of spherical coordinates in this way gives us what is termed a geographic coordinate system. Any point with respect to this axis system can be specified as (R, ϕ, λ) (Fig. 8). R is the distance of the point from the origin. The angle ϕ, the latitude, is the angle that a line passing through the point and the origin makes with the equatorial plane. The angle λ, the longitude, is the angle that a meridian passing through the point makes with the reference, and is expressed as either east or west of the reference meridian. The xz-plane is the plane of 0 degrees longitude. The xy-plane is the equatorial plane.

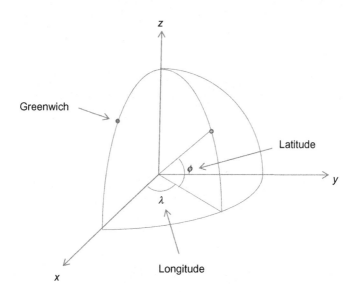

Fig. 8 Cartesian (x, y, z) and spherical (R, ϕ, λ) coordinates.

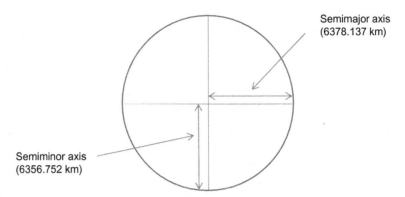

Fig. 9 Ellipsoid model of Earth (drawn approximately to scale).

A more accurate model of the Earth would be an ellipsoid of revolution, where an ellipse is rotated about its minor axis (see Fig. 9). The flattening is not visible at the scale shown. A complication is that gravity differs over the surface of the Earth. A further complication is that a position on the Earth's surface is not generally on the ellipsoid, as shown in Fig. 10.

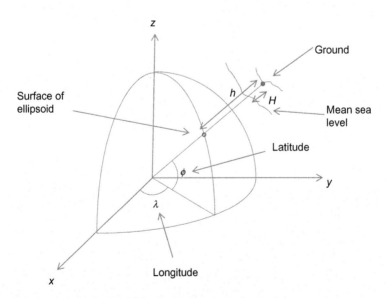

Fig. 10 A position on the Earth's surface relative to other surfaces.

When surveyors started to use GPS, then had to become aware of geodesy. The height (h) given by a GPS receiver is the height from the surface of the reference ellipsoid to the ground surface. However, the surveyor requires the orthometric height (H).

SATELLITES IN ORBIT

Prerequisites for accurately calculating the position of a GPS receiver is the receiver's ability to track satellites and receive signals broadcast from satellites, including navigation data. A receiver must have knowledge of the exact positions of the satellites. Calculating the position of a satellite is termed reference positioning.

Kepler's Laws and Orbital Dynamics

Kepler's laws describe satellite motion. However, there are factors that perturb a satellites motion.

To determine the position of a GPS receiver we need, for each satellite, its position and its distance from the receiver.

NAVIGATION SIGNALS

Navigation Signal

The signal sent by each satellite to receivers includes the satellites ephemeris data, its time, the satellite clocks correction parameters, and supporting information. From this data the receiver is able to calculate its position.

Navigation Data

Data Content

A satellites position is periodically uploaded by the GS. However, the rate at which the GS can update the position is less than the rate at which a satellite would like its new position to be known to receivers. As a result, the satellite does not broadcast its position but instead broadcast ephemeris data from which a receiver can calculate the satellites position at the start of each signal cycle. The accuracy with which these calculations are done affects the accuracy with which the receiver's position can be estimated. The orbit of a satellite is subject to perturbations. Information about the perturbations is part of a satellites signal. This data is used to get a better estimate of the receiver's position.

A satellites clock is subject to a minor drift. This effect needs to be corrected for, and the procedure for doing this is now described. The drift characteristics are calculated by the GS and transmitted to the satellite. The characteristics are then transmitted by the satellite to receivers as clock correction parameters. The actual drift amount cannot be sent when a receiver needs it as the GS updated calculations are sent too infrequently to the satellite. As a result, clock correction parameters are sent to a receiver so it can estimate the actual drift at the time it needs too. This procedure is analogous to the way in which satellite positions are calculated by the receiver from ephemeris data due to GS updates being too infrequent. The total error in a satellites clock is calculated using:

$$\Delta t_{sat} = a_0 + a_{f1}(t - t_0) + a_{f2}(t - t_0)^2$$

where a_0 is the clock offset in seconds, a_{f1} is the clock drift in second/second, and a_{f2} is the rate of clock drift in second/second^2, t_0 is a reference time, and t is the time for which one requires an estimate.

DIFFERENTIAL POSITIONING

The type of positioning described so far is called absolute positioning. In order to reduce the effect of errors for civilian navigation, and hence improve the usefulness of GPS, we can instead use differential positioning.

Differential GPS (DGPS) involves the use of several fixed ground-based reference stations. The precise location of each station is known. The actual distances from a station to each satellite are, therefore, known. Nevertheless, the station calculates the pseudodistances in the usual way and reads the navigation message. For each satellite, the difference between the actual distance to the station and the pseudodistance is calculated, thus the errors in the data are determined.

In order for a user receiver to improve its position estimate it requires additional information. For a GNSS receiver, the pseudodistances are calculated and then improved by using difference information from a reference station that is in the vicinity of user. This information is supplied in the form of a signal broadcast from a reference receiver, one that lies relatively close to the user receiver. DGPS is illustrated in Fig. 11. The idea behind DGPS is that two receivers located close to one another will both suffer from the same magnitudes of atmosphere-related errors.

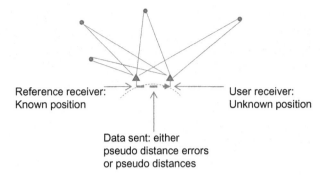

Reference receiver: User receiver:
Known position   Unknown position

Data sent: either
pseudo distance errors
or pseudo distances

Fig. 11 Differential GPS.

The magnitudes of many of the errors that contribute to the inaccuracy of GPS positioning are the same for both the reference receiver and the user receiver. Corrections can be made for ionospheric and tropospheric effects as these are almost the same at places on Earth that are located relatively close to one another. Multipath effects at the receiver cannot be corrected by augmentation as the causes of multipath are specific to the immediate location of a receiver. Another name for a reference station is a beacon. In the context of differential GPS, another name for a user receiver is a roving receiver.

DGPS can attain accuracies that match or better those achieved with the PPS. An example application area of DGPS is marine navigation.

Classifications of Differential Positioning

In differential positioning the receiver positioning errors are eliminated or significantly reduced by using the known position of the reference receiver.

Absolute/Relative Differential Positioning

With this type of differential positioning corrections are done to the pseudodistances (i.e., estimated distances from satellites to receiver) at the user receiver before a positioning calculation is made. There are two approaches. In one approach, details of pseudodistance errors are passed to the user receiver from the reference receiver. The user receiver then corrects its pseudodistances. This approach is termed absolute differential positioning.

With the second approach, the pseudodistances calculated at the reference receiver are passed to the user receiver. The user receiver then makes use of its own pseudodistances plus the pseudodistances passed to it from the reference receiver. This approach is termed relative differential positioning. Relative positioning can be used whether a user receiver is static or moving.

Real-Time/Postprocessed Differential Positioning

Real-time differential positioning requires a radio link between the user and reference receivers. It is used in applications such as real-time surveying, although it is less accurate than postprocessed differential positioning.

For high precision, one can use a simple receiver that stores GPS measurements and a reference station that logs observations. Postprocessing of the stored data takes place and corrections are then applied to the measurements stored on the receiver.

Large-Scale DGPSs

The US Coast Guard Maritime Differential GPS Service was set up with several reference stations and a couple of control centers. Its success prompted the US Department of Transportation (DOT), in 1997, to design an expansion of the Coast Guard's service to cover land areas—the continental United States as well as major transportation routes in Alaska and Hawaii.

The Process of Differencing

Differencing is a technique that simultaneously takes measurements at two receivers. There are three categories: single difference, double difference, and triple difference.

Single Difference

This is when two receivers simultaneously observe the same satellite during a single epoch (a 1.5 s period). By doing this, the errors common to both receivers cancel out. These include atmospheric delay, orbit errors, and satellite clock errors. Fig. 12 illustrates single-differencing. With single-differencing we only vary the receivers.

Double Difference

This is when two receivers simultaneously observe two (or more) satellites during a single epoch. In addition to the errors cancelled by single-differencing, double-differencing removes receiver clock errors. Fig. 13

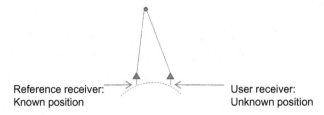

Fig. 12 Single-differencing.

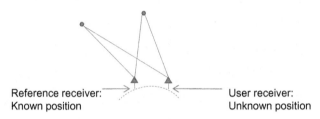

Fig. 13 Double-differencing.

illustrates double-differencing. With double-differencing we vary the receivers, and the satellites.

Triple Difference

This is when two receivers simultaneously observe two (or more) satellites during a single epoch, and then repeat the observation in the next epoch. Fig. 14 illustrates triple-differencing. With triple-differencing we vary the receivers, the satellites, and the epochs.

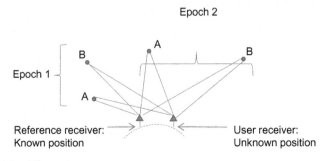

Fig. 14 Triple-differencing.

RELATIVITY

The purpose of this section is to note that relativistic effects have to be taken into account in the GPS system. Albert Einstein's equations have major implications for satellite navigation, including the GPS. Calculations relating to special relativity (SR) and general relativity (GR) effects have to be carried out.

While the ECEF frame is useful for navigation, many physical processes are easier to describe in the inertial reference frame. Following is a description of how to convert from the coordinates of a point in the initial frame to its coordinates in the ECEF frame. Let a point in the inertial frame be denoted by the cylindrical coordinates (t, r, ϕ, z). The point in ECEF is denoted by (t', r', ϕ', z'). Denote the uniform angular velocity of the Earth by ω_E. The coordinates are related to one another as follows:

$$t = t', r = r', \phi = \phi' + \omega_E t', z = z'$$

A number of SR and GR effects relate to GPS (Ashby, 2003). The velocity of a satellite clock is relatively small and the gravitational fields near the Earth are relatively weak. Both these aspects, however, cause significant relativistic effects. The fundamental concept upon which GPS is based is that the speed of light, c, is constant. The satellites contain clocks stable to about 4 ns over one day. At the speed of light, a 1 ns error is about 30 cm. If the speed of light varied then a GPS measurement would be out by 30 cm or more.

In the calculations, we need to take account of the gravitational fields near the Earth due to the Earth's own mass. The relevant expression in the amended version of the solution of Einstein's field equations involves a number of components, including the Earth's quadrupole moment coefficient and centripetal potential.

There is a relativistic effect that is dependent on the reference frame in use. It is applicable to the ECEF frame but not to the inertial frame. The effect needs to be taken into account when a satellites clock is being synchronized with one on Earth. A fixed point on the Earth's surface does not change coordinates as the Earth rotates. However, the coordinates of a satellite are changed by this rotation. Ashby (2003) gives the total time taken for a signal to travel between a satellite and a clock on Earth as:

$$\int_{path} dt' = \int_{path} \frac{d\sigma'}{c} + 2\frac{\omega_E}{c^2} \int_{path} dA'_z$$

where $(d\sigma')^2 = (dr')^2 + (r'd\phi')^2 + (dz')^2$

$$Az' = r'^2 d\phi'/2$$

The last term in the equation represents the relativistic effect. If we ignore it and try to use a satellites time to set the times on a series of clocks around the equator then, in the worst case, two Earth clocks would differ by 207.4 ns. In tests to measure the effect, two clocks positioned at different places on the Earth's surface have differed in time by as much as 350 ns. If we did not correct for this effect, then the calculated position of a GPS receiver could be in error by up to about 60 m. This would cause problems for car sat-navs. To correct for this effect, the clock receiving the signal should not be set to the time on the transmitting clock but adjusted by the amount $2\frac{\omega_E}{c^2} \int_{path} dA'_z$. The effect just described is known as the Sagnac effect.

In the 1980s, formal procedures were formulated for taking account of the Sagnac effect when comparing time standards on different parts of the Earth. The bodies responsible for these procedures are the Consultative Committee for the Definition of the Second and the International Radio Consultative Committee. If a GPS satellite is in view by two locations on Earth then the times on the locations clocks can be compared.

The other effects are smaller. The main one is the eccentricity correction. This is due to the fact that the satellites follow an elliptic rather than a purely circular orbit. With an orbit eccentricity $e = 0.01$, GR predicts a contribution of about 23 ns (i.e., an error of 7 m in the calculated position of a receiver). The eccentricity correction could be made either by the receiver or by the satellite.

A report by the Aerospace Corporation claimed that the calculations used to estimate the eccentricity correction would not be applicable in some circumstances, such as where the receiver was located in a satellite and the satellites orbit was highly eccentric. This proved to be a very controversial issue, and so in 1995 an experiment was undertaken to determine the effect of eccentricity. The TOPEX satellite was used. (Normally TOPEX is used to measure the height of the sea.) A receiver on the TOPEX satellite was used to calculate the pseudodistances. The clock in this receiver was an ordinary one and so is subject to error. The receiver was a dual frequency one and so could receive two signals broadcast from each satellite. Data was gathered by the Jet Propulsion Laboratory on October 22nd and 23rd. This included: the positions and satellite clock times of 25 GPS satellites, the positions of TOPEX, and the pseudodistances measured by the GPS receiver on TOPEX. At any one time the position of the receiver was calculated from either five or six pseudodistances. Measurements on satellite SV No. 13, having the largest eccentricity of the GPS satellites at $e = 0.01486$, showed little difference between theory and practice. If the eccentricity correction had not been taken into account, there would be an error of 10m in the calculated position of the receiver. The tests convinced GPS management to continue taking account of the effect in the next generation of satellites. During the tests, the satellite positions were measured to within 1 mm.

Ashby (2003) describes some of the other relativistic effects. Those associated with satellite clocks include the eccentricity correction, described above, and five effects which result in a satellite clocks frequency having to be set to something other than 10.23 MHz at launch. The five effects manifest themselves in that the frequency of a satellites clock as observed from the Earth's surface is greater than the actual clock frequency. As a result, a satellites clock has its frequency reduced by 4.4647×10^{-10}. Even with this correction, the frequency as observed from Earth may differ slightly from 10.23 MHz. This is due to such things as clock drift, environmental changes, and the satellite not being launched into precisely the planned orbit. In the earliest GPS satellite deployments, not all the relativistic effects associated with satellite clocks were handled correctly.

Another effect is the Doppler effect. A further relativistic effect is due to crosslink ranging. As GPS improves, relativistic effects that have hitherto been too small to be considered, are gaining attention. These include changes to a satellite clocks frequency, as observed from Earth, resulting from a change in the satellites orbit. This change in observed frequency caused by a change in orbit has been known for some time but it was only in the year 2000 that reliable measurements were taken. During one particular orbital change the frequency shift of the on-board clock was measured. In addition to the above, different types of perturbation can occur that affect the orbit of a satellite.

There are a small number of other relativistic effects that would cause an error of a few centimeters in the calculated position of a receiver (Ashby, 2003). (This equates to a 100 ps change in the time taken to transmit a signal from a satellite to a receiver.) The effects are: signal propagation delay; effect on geodetic distance (which only causes an error of a few millimeters); phase wrap-up; effect of other solar system bodies.

THE IONOSPHERE AND THE TROPOSPHERE

The ionosphere and troposphere are layers that surround the Earth. The ionosphere lies between 50 km and 1000 km above the Earth, while the troposphere lies between the Earth's surface and 80 km above it. The layers have an effect on radio waves propagated through them, that is, transionospheric and transtropospheric propagation, due to refraction. The radio waves move more slowly as they pass through them. The satellite signal delay errors worsen as a satellite moves from being directly overhead toward the horizon. Hence it is important to take this into account in GPS calculations. Further problems could be caused by solar maxima, magnetic storms, and scintillations.

IONOSPHERIC SCINTILLATION

The reduction in the strength of a GPS signal is called fading. If there is a small reduction in strength it is called shallow fading, whereas a large reduction is called deep fading. Frequent deep fading can be caused by ionospheric scintillation. It is of major concern for GPS-enabled aircraft navigation when flying in the equatorial region during the phenomenon of a solar maximum. Empirical research on the phenomenon of ionospheric scintillation has been undertaken by Seo, Walter, Chiou, & Enge (2009); this will now be described.

Seo et al.'s research involved studying data that was recorded during a solar maximum period, in which scintillation was occurring. Data collection took place at Ascension Island in the South Atlantic Ocean. A NAVSYS DSR-100 receiver with a Rubidium clock was used to gather raw data. The raw data from this receiver was then processed by a NordNav software receiver. From the final data, a period when the most severe scintillation took place was selected. This period lasted for 45 min. The signal to noise ratio of the carrier wave of a satellite signal (C/N) varied only slowly when there was no scintillation. However, during strong scintillation, C/N made frequent sudden changes, sometimes dropping by in excess of 25 dB. In the worst case, the receiver had LOS with eight satellites, seven of whose C/N values were affected by an interval of severe scintillation lasting 100 s. Deep fading occurred frequently with nearly all of the signals.

From the data, it could be concluded that during a period of severe scintillation, nearly all of the signals from those satellites in view of the receiver experienced deep fading. However, the times at which deep fading occurred differed slightly between signals.

When C/N, for a signal from a satellite, is small then the receiver loses the signal. When the C/N value increases, the signal is eventually reacquired by the receiver. The time taken for a receiver to reacquire a signal is important. If it is too long then it increases the likelihood of a receiver losing signals from several satellites, and thus affecting the accuracy of positioning or, indeed, the ability to perform positioning. (A fundamental requirement of GPS positioning is the ability to track at least four satellites.) A receiver could be designed so that it is able to cope with a short reacquisition time. In order to assist designers, this research studied the duration of signal deep fading episodes that occur during strong scintillation. The research considered the data, gathered during a 45 min period, and looked at the number of signals that would be lost by a deep fade if a receiver could cope with a 1 s reacquisition time. Similarly, the research looked at the number of lost signals for hypothetical receivers that could cope with larger reacquisition times, the largest of which was 20 s. (At the time of the research, WAAS MOPS specified aviation receivers that could cope with a 20 s reacquisition time.)

The research had to decide on the criterion by which deep fading of a GPS signal would be defined. The criterion the researchers chose was based on the NordNav software receiver that they used to gather the data. This software receiver usually lost a signal when C/N faded to below 20 dB-Hz and so the researchers decided to define deep fading, for the purposes of this research, to be whenever C/N is 20 dB-Hz or less. Although C/N fluctuated throughout the 45 min period, only deep fades were focused upon. As different receivers possess different characteristics, different definitions of deep fading may be suited to different receivers.

Seo et al.'s research developed a graphical model that relates C/N values to fading duration, for different definitions of fading. Deep fading impacts upon GPS navigation and, in the context of the research, upon aviation receiver design. A receiver could make use of a technique called coasting in its design. Using coasting, if a satellites signal is lost for a short duration, the receiver can quickly reacquire it when the C/N value becomes sufficiently large. As regards fading duration, this is defined as the time from when C/N falls below a certain threshold up to the time when C/N rises above the threshold. The researchers studied different thresholds between 15 and 30 dB-Hz. Unfortunately, from time to time, the NordNav software receiver recorder C/N values that were clearly erroneous, and such occurrences were discarded from the data considered by the researchers. The higher the C/N threshold for fading, the greater is the fading duration. The relationship between C/N threshold and fading duration was plotted for all the satellites in the 45 min period. The range of thresholds considered was then subdivided into 1 dB bins. For example, one bin was centered at 25 dB-Hz and had a 1 dB width. For each bin, the number of fades was counted. Each bin was then studied in turn to derive the distribution of fading durations. From each distribution, statistics were derived. For example, for the 25 dB-Hz bin, the median fading duration was 0.29 s and the 95th percentile was 0.88 s. For the 30 dB-Hz bin, the median fading duration and 95th percentile were longer times. Overall, median and 95th percentile values were calculated for all of the bins from 15 to 30 dB-Hz and the figures formed the basis of a fading duration graphical model. The model comprised two graphs, one of the medians and one of the 95th percentile values. Both graphs were drawn on the same set of axes: fading duration versus C/N. With the data studied, the graphs in the model were symmetrical; however, the researchers thought that in general this would not be the case. The model has multiple uses. The fading duration model is applicable to environmental conditions in which strong scintillation takes place. It is based on the finding of a certain type of software receiver. A revised model would need to be constructed for a type of software receiver, as different software receivers can behave quite differently. Furthermore, a more accurate model would require more data than was used in this research. Another limitation is that the receiver used to collect the raw data dates from 2001. It does not calculate C/N values as accurately as receivers using current technology. Nevertheless, the fading duration model presented in the research is of some use.

The time between deep fades is an important characteristic in determining whether or not a satellites signal is lost to a receiver. An aviation receiver using GPS and WAAS performs carrier smoothing and there are problems if a signal is lost during the smoothing period. Initially, when studying the time between deep fades, the research used deep fading to mean when a signals C/N falls below 20 dB-Hz. The definition of the time between deep fades used in this research was the time taken to go from the bottom of one trough to the bottom of the next trough. The distribution of the time between deep fades was studied for all of the satellites, while strong scintillation was occurring. It should be noted that for a receiver that can quickly reacquire signals, it may be able to track sufficient satellites for positioning in the event of strong scintillation.

The researchers went on to consider the effects of changing the threshold for what constitutes deep learning. Three cases were considered. One was when C/N falls below 20 dB-Hz. Another was when C/N falls below 25 dB-Hz. Obviously, the 25 dB-Hz case has more deep fades than does the 20 dB-Hz case. The third case considered was when C/N falls below 30 dB-Hz, and, as expected, there are even more deep fades. It is usually the case that the closer that the deep fade's bottom is to the threshold the shorter is the deep fade duration. Therefore, the more deep fades there are, the longer is their duration and the time between deep fades shortens. This fact is reflected in the fading duration model discussed above. In the 30 dB-Hz case, the median time between fades is 2.7 s.

The results of the research enable receiver manufacturers in their implementation of the coasting technique. The results also point to the fact that there are only short durations between deep fades and this could be problematic for aircraft that are landing, where accuracy and precision is paramount.

FURTHER READING

Kavanagh, B. & Slattery, D.K. (2014). *Surveying with construction applications* (8th ed.). London & New York: Pearson.

REFERENCES

Ashby, N. (2003). Relativity in the global positioning system. *Living Reviews in Relativity, 6(1)*, 1–42, Retrieved from http://relativity.livingreviews.org/Articles/lrr-2003-1/ (Accessed 26.04.16).

Forssell, B. (n.d.). *The dangers of GPS/GNSS. Coordinates, Feb.* Retrieved from http://mycoordinates.org/the-dangers-of-gpsgnss (Accessed 26.04.16).

Seo, J., Walter, T., Chiou, T.-Y., & Enge, P. (2009). Characteristics of deep GPS signal fading due to ionospheric scintillation for aviation receiver design. *Radio Science, 44(1)*, 1–10.

CHAPTER 4

Signals From Satellites to Receiver—GPS

Signal processing is an important area in which mathematical concepts and techniques can be applied. Its importance has grown due to developments in big data processing, genomics and bioengineering, mobile communications, multimedia systems, networks, neural signal processing, and the internet of things.

Fig. 1 is a block diagram that represents the relevant communications link.

GPS SIGNAL STRUCTURE

We have digital data. Some transmission media, such as optical fiber and unguided media (such as the atmosphere), will only propagate analog signals. Therefore, we wish to transmit digital data using an analog signal. Another example of transmitting digital data using an analog signal is when computers are communicating using voice (over the plain old telephone system). Modulation of the analog signal is performed. Modulation is the varying of one or more of the properties of a periodic waveform, called the carrier signal. Three encoding techniques are used to transmit digital data using analog signals—amplitude-shift keying (ASK), frequency-shift keying (FSK), and phase-shift keying (PSK). Keying means the modulating signal takes one of a specific number of values at all times.

Carrier Signal

The satellites continuously broadcast electromagnetic signals in the L-band, which is used for radio communication.

Each satellite transmits two sinusoidal carrier signals, L1 and L2, with frequencies of 1575.42 and 1227.60 MHz, respectively. These frequencies

Uncertainties in GPS Positioning
http://dx.doi.org/10.1016/B978-0-12-809594-2.00004-6

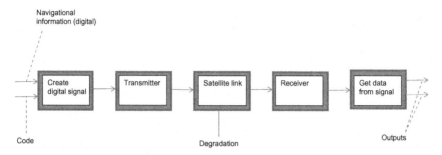

Fig. 1 Block diagram of the communications link.

are coherently selected integer multiples of the fundamental clock rate, or frequency, $f_0 = 10.23\,\text{MHz}$:

$$L1 = 154 \times f_0$$
$$L2 = 120 \times f_0$$

The relationship between the wavelength (λ), the speed of light (c), and the frequency (f) is $\lambda = c/f$. For L1:

$$\lambda = \frac{300,000,000\,\text{m/s}}{1,575,420,000\,\text{Hz}} = 0.190\,\text{m},$$

that is, the wavelength is about 19 cm. For L2, the wavelength is 0.224 m; about 24 cm.

GPS uses code division multiplexing, enabling all of the satellites to use the same carrier frequency without the signals interfering with one another. The L1 carrier is modulated by three binary sequences—a pseudorandom noise (PRN) bit sequence conforming to the coarse/acquisition (C/A) code, a PRN conforming to the P(Y) code, and the navigation message. The L2 carrier is modulated by two binary sequences—a PRN conforming to the P(Y) code and the navigation message. A carrier signal is modulated using binary phase-shift keying (BPSK). With PSK, the phase of the carrier is shifted to represent data, as shown in Fig. 2. There does not have to be one period of the wave for each bit. There could be multiple periods per bit, for example, two periods per bit, two and a half periods per bit, etc.

Exercises

Sketch the signals in Tables 1 and 2.

Fig. 3 shows the phase states corresponding to the PSK waveform in Fig. 2.

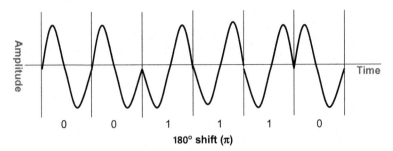

Fig. 2 Phase-shift keying.

Table 1 First exercise for PSK

Encoding technique		0	1	0	0	0	1	1	1	0	0	0	0	0	0	0	0	1	1	1	1	1
PSK	+0−																						

Table 2 Second exercise for PSK

Encoding technique		1	0	0	0	0	0	0	0	0	1	0	1	0	1	1	0	0	0	0	1	1
PSK	+0−																						

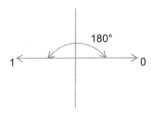

Fig. 3 Phase states.

Let us now consider a more general case, where the number 0 is represented by a part of a cosine wave with phase θ_0 and the number 1 is represented by a cosine wave with phase θ_1. For the number 0, we have:

$$s_0(t) = V_c cos(\omega_c t + \theta_0) \quad 0 < t \leq T$$

where V_c is the carrier amplitude, ω_c its angular frequency, t is time, θ_0 is a constant phase shift, and T is the time taken to represent a single bit.

For the number 1, we have:

$$s_1(t) = V_c cos(\omega_c t + \theta_1) \quad 0 < t \leq T$$

where θ_1 is a constant phase shift.

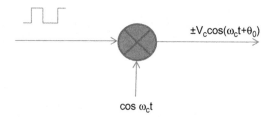

Fig. 4 Modulating the phase of a carrier.

Table 3 Signal element represents two bits

Bit value	Amount of shift
00	None
01	1/4 of a period
10	1/2 of a period
11	3/4 of a period

For a phase difference of 180 degrees, that is, where $\theta_1 = \theta_0 + 180$ degrees, we have:

$$s_1(t) = -s_0(t) = -V_c cos(\omega_c t + \theta_0)$$

Fig. 4 shows another representation of the modulation of a carrier by a digital signal, using PSK.

A more efficient usage is for each signal element to represent more than one bit, for example, two bits per time interval, as shown in Table 3.

Fig. 5 is a block diagram showing the steps involved in taking a satellite's raw digital data, transmitting it, and recreating the data at the receiver.

The L1 carrier is modulated by a combination of a PRN conforming to the C/A code and the navigation message, and then further modulated by a combination of a PRN conforming to the P(Y) code and the navigation

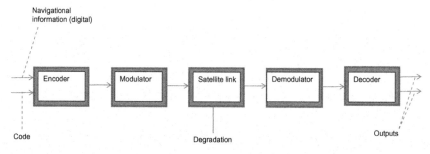

Fig. 5 Block diagram of the coding system.

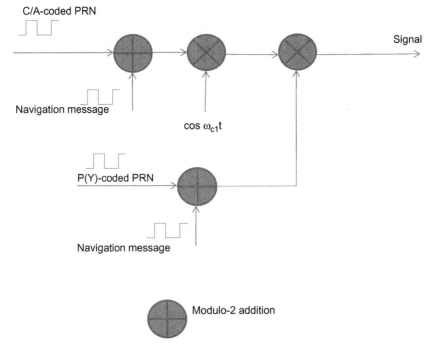

Fig. 6 Modulation of the L1 carrier.

message, as shown in Fig. 6. The L2 carrier is modulated by a combination of a PRN conforming to the P(Y) code and the navigation message, as shown in Fig. 7.

C/A and P(Y) Codes

Two kinds of code are described here—a C/A code and a P (precise) code. The C/A code is a type of Gold code. A Gold code is a code for a binary

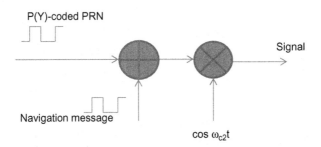

Fig. 7 Modulation of the L2 carrier.

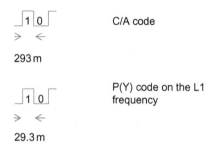

Fig. 8 Physical distance occupied by 1 chip.

sequence. Such codes are used because multiple satellites are transmitting to each receiver using the same frequency. Each satellite transmits codes that are unique to that satellite. C/A specifies that the PRN has a period of 1023 chips (i.e., pulses) transmitted at 1.023 Mbit/s (equal to $f_0/10$). This is repeated every millisecond. The C/A code is freely available to the public. As regards the P(Y) code, it specifies that the PRN is to be transmitted at 10.23 Mbit/s (equal to f_0) on the L1 frequency and 0.5115 Mbit/s on the L2 frequency. The physical distance that each code occupies as it travels from a satellite is shown in Fig. 8.

Both a PRN conforming to the C/A code and a PRN conforming to the P(Y) code are transmitted on the L1 frequency, whereas only a PRN conforming to the P(Y) code is transmitted on the L2 frequency. The P(Y) code specifies a much longer PRN than does the C/A code. Using the P(Y) code gives more accurate positioning. PRNs conforming to the P(Y) code are only accessible to users using the PPS service; those using the SPS service have to access PRNs conforming to the C/A code. GPS receivers for PPS can determine position using PRNs conforming to either the C/A code, or the P(Y) code, or both.

PRNs conforming to the P(Y) code are reset every Saturday/Sunday midnight.

Table 4 shows codes applied to the carrier signals.

Table 4 C/A- and/or P(Y)-coded PRNs modulating carrier signals

		C/ACode($f_0/10$)	P(Y)Code(f_0)
Carrier	L1($154f_0$)	✓	✓
	L2($120f_0$)		✓

The carrier signals L1 and L2 are each single frequencies and so each would be represented as a vertical line on a plot in the frequency domain. However, when they are modulated using BPSK, the graphs look quite different. Fig. 9 shows the modulated L1 signal. The shape of the PRN conforming to the C/A code has a width that is twice its chipping frequency, that is, 2.046 MHz. The main lobe of the shape of the PRN conforming to the P(Y) code has a width that is twice its chipping frequency, that is, 20.46 MHz. Fig. 10 shows the modulated L2 signal. The width of the main lobe is also 20.46 MHz.

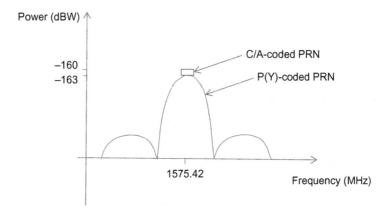

Fig. 9 The modulated L1 carrier in the frequency domain.

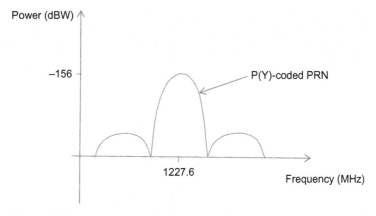

Fig. 10 The modulated L2 carrier in the frequency domain.

Data

A GPS satellite transmits the time at which a transmission is sent and the satellite's position. This information can be used for navigation purposes. There are also other items of data transmitted such as the satellite's health, the satellite's clock correction, and the propagation delay effects.

Fig. 11 shows the navigation message data structure.

A GPS receiver can obtain the almanac data from any satellite. It tells the receiver where in the sky each satellite will be located at any given time.

Fig. 12 illustrates for the C/A ranging code what happens in one second of signal transmission.

Navigation Processing

The method used to calculate a pseudodistance is now described. The method is similar to that used in other GNSSs.

Subframe-1	TLM	HOW	Satellite clock and health data (satellite clock correction parameters, etc.)
Subframe-2	TLM	HOW	Satellite ephemeris data
Subframe-3	TLM	HOW	
Subframe-4	TLM	HOW	Support data (ionospheric and UTC data, etc.)
Subframe-5	TLM	HOW	Support data (almanac data for satellites 1 to 24, etc.)

Subcommuted 25 times each

Each subframe has 300 bits and takes 6 seconds to transmit

Fig. 11 Navigation message data structure.

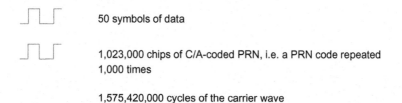

50 symbols of data

1,023,000 chips of C/A-coded PRN, i.e. a PRN code repeated 1,000 times

1,575,420,000 cycles of the carrier wave

Fig. 12 What gets transmitted in 1 s.

The time used by GPS is with respect to the GPS time scale. The time of transmission is repeatedly transmitted by a satellite and is used in pseudodistance calculations. The navigation message is comprised of frames, each frame being made up of subframes. Each subframe starts with a TLM word (telemetry) followed by a HOW word (handover). TLM comprises an 8-bit preamble, followed by 16 reserved bits, followed by 6 parity bits. HOW comprises:

a. Seventeen bits called the truncated time of week (TOW) count. This shows how many 6 s periods have elapsed since the full TOW count was reset. Resetting occurs at midnight Saturday night–Sunday morning.

b. One bit used as either a momentum flag or an alert flag.

c. One bit used as either a synchronization flag or an antispoofing flag. In the case when bit 19 is used as a synchronization flag, the value of the flag indicates whether or not the satellite has been synchronized with the constellation. If it has, the 1.5 s taken to transmit a TLM word is coincident with the constellation's 1.5 s epoch.

d. Two bits to represent the subframe ID.

e. Two bits used following computation involving the parity bits.

f. Six parity bits.

HOW is essential to pseudodistance calculations.

The GPS time scale has been set up with reference to UTC, which is a scale maintained by the US Naval Observatory. Zero on the scale (called the zero time-point) corresponds with midnight on the night of January 5th/morning of January 6th, 1980. The largest period of time referred to in the GPS time scale is the week, lasting 604,800 s. Let us consider how GPS time is stored and how it relates to UTC time. GPS time makes use of a 29-bit Z-count. The Z-count comprises two parts. The 19 least significant bits (LSBs) represent the TOW-count, which is the number of 1.5 s epochs that have occurred since these bits were reset. Resetting occurs at midnight Saturday night–Sunday morning. The number of 1.5 s epochs in a week is $604,800/1.5 = 403,200$. The LSBs, therefore, range from decimal 1 to decimal 403,199. The 10 most significant bits (MSBs) record the number of weeks that have elapsed in GPS time since midnight on the night of January 5th/morning of January 6th, 1980 (modulo 1024). The complete Z-count identifies a specific 1.5 s epoch in the last 20 years, or so.

At the end of a GPS week the LSBs are reset and the MSBs are incremented.

Each subframe of the navigation message takes 6 s to transmit. The truncated TOW-count is part of a subframe's HOW word, and the

TOW-count shows the time that the subframe was transmitted. Therefore, the TOW-counts for adjacent subframes differ by 1, meaning 6 s.

Consider a PRN that conforms to the C/A code being transmitted from one of the block IIA or block IIR satellites. The time to transmit a subframe of the navigation message is:

No. of words × No. of bits per word/bit rate $= 10 \times 30/50 = 6\,\text{s}$

In six seconds, the number of PRNs that can be transmitted is:

(PRN bit rate/No. of bits in a PRN) $\times 6 = (1{,}023{,}000/1023) \times 6 = 1000$

All satellites in the constellation are synchronized and transmit the navigation messages at the same time. The HOW word of each subframe contains the truncated TOW-count which specifies the time at which the next subframe will start being transmitted.

FURTHER READING

Calcutt, D., & Tetley, L. (1994). *Satellite communications: Principles and applications.* Oxford: Butterworth-Heinemann.

CHAPTER 5

GPS Modernization

In 1999, the United States announced its intention to modernize the GPS system and develop a new generation of satellites that broadcast two additional civilian signals. Currently there are two GPS signals, one at the L1 frequency (1575.42 MHz) and one at the L2 frequency (1227.60 MHz). A major focus of modernization is the provision of new ranging codes (Jung, Enge, & Pervan, n.d.). Currently new civil ranging codes are under development. These are L2C, L5, and L1C. The legacy code C/A will continue. L2C is the second civilian code. When a PRN conforming to C/A is combined with a PRN conforming to L2C in a dual-frequency receiver, ionospheric correction is enabled. L2C delivers faster signal acquisition, enhanced reliability, and greater operating range. One of its features is a dedicated channel for codeless tracking. L2C broadcasts at a higher effective power than C/A, which makes the signals easier to receive under trees and indoors.

L5 is the third civilian ranging code and is to be used for safety-of-life transportation and other high-performance applications. It gives a higher-power signal, with greater bandwidth, and with a new design. The greater bandwidth improves jam resistance. The modern signal design offers multiple message types and forward error correction.

L1C is the fourth civilian ranging code and is designed to be interoperable with other GNSSs. A feature of L1C is its Multiplexed Binary Carrier Offset modulation scheme. Better mobile reception is envisaged in cities and other difficult environments. Once again, the design includes forward error correction.

GPS uses more than its original 24 satellites. The constellation is made up of both old and new ones. There are different generations (termed blocks) of satellites. When GPS became operational block II satellites were used. Today, the legacy satellites belong to block IIA and block IIR. The modernized satellites begin with block IIR(M), to which the L2C ranging

Table 1 GPS satellite constellation

Plane	Number of satellites	Blocks
A	4	2 IIR(M); 2 IIF
B	5	2 IIR; 1 IIR(M); 2 IIF
C	5	1 IIR; 2 IIR(M); 2 IIF
D	5	3 IIR; 2 IIF
E	6	2 IIR; 2 IIR(M); 1 IIF; 1 IIA
F	5	3 IIR; 1 IIR(M); 1 IIF

code was added. More recently are the block IIF satellites, to which the L5 ranging code was added, and which has a longer design lifespan. A block referred to as GPS III is soon to be launched, and it will include the L1C ranging code.

Table 1 shows the satellite constellation at the time of writing (20th Jan. 2016). Block IIA satellites were launched between 1990 and 1997; block IIR were launched between 1997 and 2004; block IIR-M were launched between 2005 and 2009; block IIF have been launched since 2010; block III are in production and will be available for launch in 2016. Each satellite circles the Earth twice daily. The satellites are arranged on six equally spaced orbital planes.

FREQUENCY PLAN

GPS satellites transmit on two frequencies—L1 (1575.42 MHz) and L2 (1227.60 MHz). In the future, there will be a third frequency—L5 (1176.45 MHz). Satellite blocks IIA, IIR, and IIR(M) use both L1 and L2. Block IIF uses L1, L2, and L5 and so will block III satellites. In block IIA and IIR, the L2 frequency was only used as part of the precise position service. The frequencies are based on a fundamental frequency of 10.23 MHz:

$$L1 = 154 \times 10.23, L2 = 120 \times 10.23, L5 = 115 \times 10.23$$

The actual frequency used in a satellite is 10.22999999543 MHz. The reason for this is to compensate for relativistic effects, so that the satellite's frequency when viewed from Earth appears to be 10.23 MHz.

The velocity of the signals transmitted from a satellite varies as the satellite approaches the receiver and then as it recedes from the receiver. This phenomenon is known as the Doppler effect. The change in velocity of transmission is directly proportional to the shift in frequency of the transmitted signals, and is also proportional to the change in distance between the satellite and the receiver over a specified time interval.

A signal transmitted by a satellite with an L1 frequency is received at a receiver with a different frequency due to the Doppler effect, caused by the satellite moving with respect to the receiver. Similarly, a signal transmitted with an L2 frequency is not received at that frequency, and a signal transmitted with an L5 frequency is not received at that frequency. For a stationary receiver on the Earth's surface, the Doppler effect makes a change of up to 5 kHz. For a moving receiver on the Earth's surface, the change can be greater. GPS receivers compensate for these frequency shifts.

The data (i.e., the navigation message) that a satellite wishes to transmit is digital. Many satellites are communicating at the same time. There are different techniques that could be used so that the receiver can decompose what it is receiving into distinct satellite signals. The technique used in GPS is called CDMA. It makes use of pseudorandom noise (PRN) bit patterns that are unique to each satellite. The data and the PRN(s) are combined and sent as an analog signal. The PRN conforms to one of a number of ranging codes, such as the C/A code and the P(Y) code. The choice of a ranging code affects the characteristics of the PRN, the data, and the analog signal. The ranging codes are given in Table 2. In order to distinguish data bits from PRN bits, the latter are called chips. A satellite transmits more than one analog signal, each having one or more PRNs conforming to different ranging codes. Currently, on the L1 frequency, both the SPS and PPS services are offered, whereas on the L2 frequency only the PPS service is offered. From the launch of satellite blocks IIR-M, the satellites have been testing two further ranging codes: L2 civil-moderate (L2 CM) and L2 civil-long (L2 CL). The ranging codes L2C, L5, and L1C are a major focus of GPS's modernization program. All of these new ranging codes are for civilian use. The L1 ranging code is also used in satellite-based augmentation systems (SBASs). Such a system is used in civil aviation. The GPS signals are augmented with those from geostationary satellites using the L1 ranging code.

MULTIPLE ACCESS

All satellites use the same frequency bands. The technique used so that a receiver can decompose what it receives into individual signals is code division multiple access (CDMA), and it involves each satellite in the constellation

Table 2 Ranging codes

		Ranging code(s)				
		C/A, L1 P(Y)	L2 P(Y)	L2 CM, L2 CL	L5	L1C$_{Pi}(t)$, L1C$_{Di}(t)$
PRN	Length (c)	1023 (C/A), 6.19 × 10^{12} (P(Y))	10^{12}	10,230 (CM), 767,250 (CL)	10,230	10,230
	Chip rate (Mcps)	1.023 (C/A), 10.23 (P(Y))	0.5115	0.5115	10.23	1.023
Data (navigation message)	Bit rate (bps)	50 (satellite blocks IIA, IIR); 25 (satellite blocks IIR–M, IIF, III)	50 (satellite blocks IIA, IIR); 25 (satellite blocks IIR–M, IIF, III)	25 (only satellite blocks IIR–M, IIF, III)	50	46.39
	Symbol rate (sps)	50	50	50	100	100
Analog signal	Carrier frequency (MHz)	1575.42	1227.60	1227.60	1176.45	1575.42
	Frequency band	L1	L2	L2	L5	L1
	Modulation	BPSK	BPSK	BPSK	BPSK	BOC
	Bandwidth (MHz)	20.46 (satellite blocks IIA, IIR, IIR–M, IIF); 30.69 (satellite block III)	20.46 (satellite blocks IIA, IIR, IIR–M, IIF); 30.69 (satellite block III)	20.46 (satellite blocks IIA, IIR, IIR–M, IIF); 30.69 (satellite block III)	24	30.69
Services		SPS (C/A), PPS (P(Y))	PPS	SPS	SPS	SPS

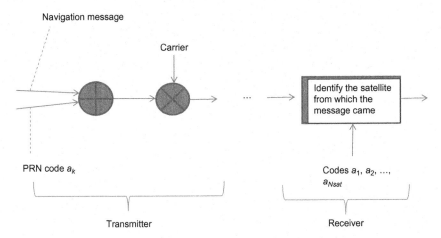

Fig. 1 CDMA technique.

being assigned a different PRN. These bit patterns are orthogonal codes. For N_{sat} satellites, the codes are a_k, where $k = 1, 2, \ldots, N_{sat}$. Fig. 1 shows how the CDMA technique works; the signal corresponding to the L2 carrier has been chosen for simplicity. The data transmitted by a satellite is termed the navigation message, and is used by the receiver to perform ranging. The digital signal in a satellite is a combination of the data and PRN, using modulo-2 addition. The data travels at a low bit rate whereas PRN travels at a high rate (chips per second). The digital signal is then converted to an analog one. This digital to analog conversion involves modulating a carrier wave. The modulation technique used is binary phase-shift keying (BPSK), for all ranging codes except for the L1C ranging code, which uses binary offset carrier (BOC). A PRN conforms to one of the ranging codes: C/A is the coarse/acquisition ranging code which is used on the L1 frequency band for civil users; P(Y) which is the precise code use on the L1 and L2 frequency bands for military users; L2C on the L2 frequency band for civil users; L5 on the L5 frequency band for civil users; L1C on the L1 frequency band for civil users. The P(Y) code is so named because one of two codes is used: the P-code when the antispoofing (A–S) mode of operation is not activated; the encrypted Y-code when A–S is activated. A–S is a technique that prevents miscreants from broadcasting imitation GPS signals. Table 3 summarizes the effects of A–S.

Table 3 The effect of A-S

	Effect on C/A and P(Y) codes	Effect on PPS users	Effect on SPS users
A-S off	Normal	Normal accuracy; spoofable	Normal accuracy; spoofable
A-S on	Normal	Normal accuracy; not spoofable	Normal accuracy; spoofable

C/A, L1 P(Y)

Let us consider the analysis of an analog signal transmitted from the antenna of a satellite. Fig. 2 shows a plot of amplitude versus frequency. The amplitude of a signal is related to the signal's power. The hill in the middle is called the main lobe. The lower hills are called the side lobes. If we transmit a main lobe and side lobes from the antenna then physically the central part of the main lobe is a main beam where the signal is strongest, and so is best suited for receivers. On the outlying parts of the main lobe, and on the side lobes, the signal is weaker but some receivers may still be able to pick up the signal. For the L1 P(Y) ranging code, the bandwidth of the central part of the main lobe is approximately 20 MHz. For satellite blocks IIA, IIR, IIR-M, and IIF the signal is transmitted at 20.46 MHz and so only the central part of the main lobe is transmitted. For satellite block III the signal will be transmitted at 30.69 MHz and so the main lobe and some side lobes will be transmitted. For the C/A ranging code, the bandwidth of the central part of the main lobe is only about 2 MHz, and so all satellite blocks transmit the main lobe and some side lobes. For the C/A ranging code, the PRN has length 1023 chips and the chip rate is 1.023 Mcps. This means that the

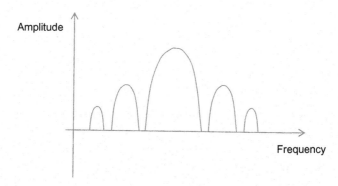

Fig. 2 Signal transmitted from a satellite antenna.

PRN repeats itself every 1 ms. A receiver has a copy of all the PRNs of the satellite constellation. As the PRN is so short, a receiver can correlate a satellite's PRN with one of those patterns that it has stored. As the PRN repeats itself every 1 ms, the receiver takes at least 1 ms to know where the PRN starts. C/A is a Gold code. These are used in CDMA and the PRN has a length of $2^n - 1$. The PRN conforming to the C/A ranging code has a length of $1023 = 2^{10} - 1$.

The bit rate of the data (i.e., navigation message) is the same when using either the C/A or L1 P(Y) ranging codes—50 bps. As no bits are added for error checking, the symbol rate is 50 sps. With satellite blocks IIA and IIR, the combination of one data bit and 20 copies of the relevant PRN take 20 ms to transmit. With satellite blocks IIR-M, IIF and III, the data rate has been halved due to the inclusion of parity bits. The navigation message comprises 25 frames. Each frame has five subframes. Each subframe consists of 10 words. Each word has 30 bits. The MSB of word is transmitted first. With satellite blocks IIA and IIR, the navigation message takes 12.5 min to transmit, whereas with satellite blocks IIR-M, III, and III it takes 25 min.

L2 P(Y)

The composite signal transmitted from satellite i is given by:

$$s_i(t) = s_{L1,i}(t) + s_{L2,i}(t) + s_{L5,i}(t)$$

The three terms on the right-hand side correspond to the three frequencies used. We can decompose each of these three terms. For example, for frequency L2 we have:

$$s_{L2,i}(t) = \text{(term applicable to ranging code L2 P(Y))}$$
$$+\text{(term applicable to ranging code L2C)}$$

The term applicable to ranging code L2 P(Y), for example, is

$$\sqrt{2}P(P_i(t) \text{ MOD2 } D_i(t))sin(2\pi f_{L2}t)$$

where P is the signal power at frequency L2, $P_i(t)$ is the PRN, $D_i(t)$ is the data bit pattern, and f_{L2} is the L2 frequency. The MOD2 symbol denotes the modulo-2 addition operation.

IMPROVED PSEUDODISTANCE CALCULATIONS

When multiple civil signals become available it will be possible for very accurate pseudodistance calculations to be made (Jung et al., n.d.).

GPS RECEIVER

It takes time for a satellite's signal to arrive at the receiver. If we were to compare the transmitted signal to the received signal, time-wise, the PRN and the navigation data are obviously shifted on the received signal. Furthermore, the frequency of the signal sent by the satellite is different to that received at the receiver due to the Doppler effect.

Fig. 3 is a block diagram to illustrate the basic functionality of a GPS receiver. The receiver receives signals from the satellites with which it has line of sight. For each of these satellites, it must track the signal, identify the PRNs, and read the navigation message.

Part of the circuitry is a phase locked loop (PLL). This is used for extracting the navigation message from the signal. The receiver decides which satellites in the constellation it is able to track. Another part of the circuitry is the carrier tracking loop. This involves the receiver generating the L1 carrier frequency. The receiver can then compare the frequency with the frequency of an L1 carrier-based signal transmitted by a satellite. The frequencies will not be the same due to the Doppler effect. The difference in frequencies is used to calculate the difference in velocity between the receiver and the satellite. The receiver also generates PRNs, which are replicas of those that the satellites in the constellation have. It matches a PRN of an incoming signal with one of the replicas. In order to do this it needs to do two things. Firstly, it must adjust the frequency of the replica so that it is the same as the central frequency of the signal. Secondly, it must adjust the phase of the replica so that it is aligned with the incoming PRN.

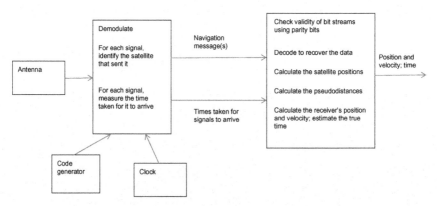

Fig. 3 Block diagram of receiver functionality.

When the receiver has track the signals of four satellites, identified their respective PRNs, and read the navigation messages, it calculates the pseudodistances.

A receiver changes the frequency of each incoming analog signal into what is termed an intermediate frequency. It then converts each analog signal to a digital signal. The nth sample of the digital signal coming from satellite k is given by:

$$S_{L1}^{(k)}(n) = \sqrt{P_c}C^{(k)}(n)D^{(k)}(n)cos(\omega_{IF}n) + e(n)$$

where P_c is the power of the signal, $C^{(k)}(n)$ is the sample of the PRN, $D^{(k)}(n)$ is the sample of the navigation data, and ω_{IF} is the intermediate frequency.

Signal Acquisition

A GNSS signal is acquired only when the frequency of the local carrier replica matches the frequency of the carrier in the received signal, and the PRN code replica is closely aligned, in time, with the PRN code in the received signal. The frequency of the carrier in the received signal is shifted due to the Doppler effect. Furthermore, the receiver's oscillator does not give a precise frequency due to the inexpensive substance used. The whole frequency search band is divided into frequency bins. Initially, only a coarse value of the carrier frequency is found and only a coarse value of the C/A code phase is found. A particular receiver may refine these values. Different acquisition techniques are in use. A straightforward method of acquisition is to search all possible combinations of code phases and carrier frequency bins. There are 41 different carrier frequency bins and 2046 different C/A code phases. The total number of combinations to search is 41 × 1023 × 2/2 = 41,943 combinations (bins).

Signal Tracking

Having acquired the received GNSS signal, it must then be tracked. Only by doing this can the navigation messages be demodulated and a precise measurement of the propagation delays be made. Tracking requires the receiver to maintain local replicas that closely follow each received satellite signal. Carrier- and code-phase tracking are carried out by feedback tracking loops. In a GNSS receiver, two feedback tracking loops are typically used for each signal—one for the carrier and one for the PRN code. The carrier is tracked by a frequency locked loop (FLL) or a PLL. The PRN code of the received signal is tracked by a delay locked loop (DLL).

The carrier-phase mismatch between the received signal and the local replica is typically measured with a carrier-phase discriminator (also known as an error detector).

Code Delay Tracking: The DLL

Let us consider a PRN conforming to the C/A code. The chip rate is 1.023 Mcps. The time to transmit one chip, T_c, is approximately 1 μs. In 1 μs, the chip has travelled a distance of c. T_c, which is approximately 300 m, where c is the speed of light. The part of the circuitry that measures how long a signal takes to travel from the relevant satellite to the receiver is the DLL. Consider a DLL that can estimate signal delay to an accuracy of 1/100th chip. The receiver is, therefore, able to measure the distance between the satellite and the receiver to an accuracy of approximately 3 m. There are 1023 chips in the PRN and so a PRN takes the satellite 1 ms to transmit. Every 1 ms the same PRN is transmitted. If the receiver makes a mistake and instead of noting the arrival time of the xth PRN, it notes the arrival time of the $(x + 1)$th PRN then the measured distance between the satellite and the receiver will be wrong by approximately 300 km. The navigation message can be used to detect when this happens. The above analysis has assumed that the satellite clocks and the receiver clocks are all synchronized. While the satellite clocks are synchronized, and are highly accurate atomic clocks, the receiver clock is not as it is a relatively inexpensive clock. As a result of this, the distance calculated by the DLL is referred to as a pseudodistance. The error in the receiver clock's time, called the receiver clock offset, is obviously the same for every satellite's signal, hence the position of the receiver and the receiver clock offset are calculated together in one algorithm.

Code tracking is often implemented as a DLL in which three replicas are generated and correlated with the incoming PRN code. The three replicas are referred to as the early, prompt and late replica. The early replica differs from the prompt replica by half a chip length. Similarly, the late replica differs from the prompt replica by half a chip length, in the opposite sense.

REFERENCE

Jung, J., Enge, P., & Pervan, B. (n.d.). *Optimization of cascade integer resolution with three civil GPS frequencies*. Retrieved from http://gps.stanford.edu/papers/Jung_IONGPS_2000. pdf (Accessed 26.04.16).

CHAPTER 6

Signals From Satellites to Receiver—Other Satellite Navigation Systems

GALILEO SIGNAL STRUCTURE

The services that Galileo expects to provide include:

1. Open service (OS): free to use; for positioning and timing.
2. Safety-of-life service: warns a user when it is unable to give accurate information. It is hoped to provide a service guarantee.
3. Commercial service (CS): includes two signals; higher data rate; improved accuracy; encrypted.
4. Public regulated service (PRS): for specific users, that is, those requiring a high continuity of service with controlled access; includes two signals; encrypted PRN bit patterns and navigation message.
5. Support for search and rescue. This service is Europe's contribution to the international COSPAS-SARSAT search and rescue effort. Galileo is to play an important part in the Medium Earth Orbit Search and Rescue (MEOSAR) system. Galileo will be able to receive signals transmitted from emergency beacons on ships, planes, or in the possession of individuals. Galileo will forward such a signal to the relevant national rescue center, enabling the center to locate the beacon. As it is hoped that every location on Earth will have LOC to at least one Galileo satellite, it is hoped to handle beacon transmission almost instantaneously. In certain circumstances, Galileo could respond to the transmitting beacon, a feature that is unique to Galileo.

Frequency Plan

The E1 signal is used in the following services: OS, CS, and PRS. It is transmitted at the L1 frequency. E1 has three component signals: E1-A, E1-B, and E1-C. For OS and CS, E1-B and E1C are used; E1-B carries data while E1-C, which is referred to as a pilot component, does not carry

Table 1 Details of E1 signal

Service			OS		PRS
Signal component			**Data**	**Pilot**	**Data**
PRN	Length (c)		4092	4092 (primary), 25 (secondary)	
	Chip rate (Mcps)		1.023		10.23
Data (navigation message)	Bit rate (bps)		125		
	Symbol rate (sps)		250		
Analog signal	Carrier frequency (MHz)		1575.42		
	Subcarrier frequency (MHz)		1.023 and 6.138 (two subcarriers)		15.345
	Frequency band		1559–1591 MHz		
	Modulation		CBOC(6,1,1/11)		$BOC_{cos}(15,2.5)$
	Bandwidth (MHz)		24.552		

data. Both E1-B and E1-C make use of ranging codes, and they do not require encryption to be carried out. For PRS, E1-A is used. The access is restricted. The data to be transmitted is encrypted and the ranging code also requires encryption to be carried out.

Table 1 gives details of the E1 signal. Tables 2 and 3 give details of the other Galileo signals.

Galileo E1

The band used by the E1 signal ranges from 1559 to 1591 MHz. The carrier frequency is 1575.420 MHz. Fig. 1 shows how the E1 signal is formed. The overall process is called Composite Binary Offset Carrier modulation. With Galileo, different navigation data streams are transmitted. The one used in an E1 signal is the I/NAV navigation data stream. Linear combinations of square waves, $s_1(t)$, $s_2(t)$, $s_3(t)$, and $s_4(t)$, are used. Each square wave is

Table 2 Details of E5 signal

Signal		E5a		E5b	
Signal component		**Data**	**Pilot**	**Data**	**Pilot**
PRN	Length (c)	10,230 (primary), 20 (secondary)	10,230 (primary), 100 (secondary)	10,230 (primary), 4 (secondary)	10,230 (primary), 100 (secondary)
	Chip rate (Mcps)	10.23			
Data (navigation message)	Bit rate (bps)	25		125	
	Symbol rate (sps)	50		250	
Analog signal	Carrier frequency (MHz)	E5 (1191.795); E5a (1176.45); E5b (1207.14)			
	Subcarrier frequency (MHz)	15.345			
	Frequency band	1164–1191.795 MHz		1191.795–1215 MHz	
	Modulation	AltBOC(15,10)			
	Bandwidth (MHz)	E5 (51.15); E5a (20.460); E5b (20.460)			

called a subcarrier. The waves $s_1(t)$ and $s_2(t)$ are in-phase while $s_3(t)$) and $s_4(t)$ are antiphase.

$$s_1(t) = sgn[sin(2\pi\ 1.023\ t)]$$
$$s_2(t) = sgn[sin(2\pi\ 6.138\ t)]$$
$$s_3(t) = sgn[sin(2\pi\ 1.023\ t)]$$
$$s_4(t) = sgn[sin(2\pi\ 6.138\ t)]$$

The values of α and β are chosen such that:

$2 \times$ [Power of $s_2(t)$ + Power of $s_2(t)$] $= \frac{1}{11}$[Power of $e(t)$ + Power of $(\alpha s_1(t) - \beta s_2(t))$], before application of any bandwidth limitation. This equates to $\alpha = \sqrt{\frac{10}{11}}$ and $\beta = \sqrt{\frac{1}{11}}$. The expression for $s_2(t)$ includes the factor 6.138×10^6 which is six times the factor 1.023×10^6 used in the expression for $s_1(t)$; furthermore there is a factor of $1/11$ used in the

Table 3 Details of E6 signal

Service			CS		PRS
Signal component			**Data**	**Pilot**	
PRN	Length (c)		5115	5115 (primary), 100 (secondary)	
	Chip rate (Mcps)		5.115		
Data (navigation message)	Bit rate (bps)				
	Symbol rate (sps)		1000		
Analog signal	Carrier frequency (MHz)		1278.75		
	Subcarrier frequency (MHz)				10.23
	Frequency band		1260–1300 MHz		
	Modulation		BPSK	BPSK	$BOC_{cos}(10,5)$
	Bandwidth (MHz)		40.92		

equation to calculate the values of α and β. The modulation used to create the E1 signal is referred to as CBOC (1,6,1/11). In summary, the carrier wave is modulated by two components signals, each of which has a 50% power sharing.

GLONASS

GLONASS stands for Globalnaya Navigatsionnaya Sputnikovaya Sistema. Originally it purpose was for military use throughout the world.

The full constellation of 24 satellites has been available since 2011. GLONASS has an accuracy comparable to GPS.

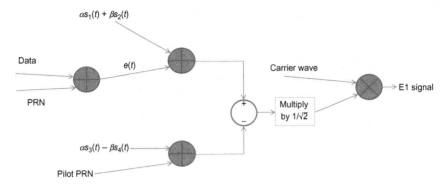

Fig. 1 How the E1 signal is formed.

System Characteristics
GLONASS Ground Segment
As with GPS and Galileo, GLONASS has a ground segment. GPS and Galileo stations are spread across the world, which helps in controlling the respective constellation and its signals. GLONASS, on the other hand, only has stations in areas that were formally covered by the old Soviet Union plus one in Brazil. Spanning more of the world would improve accuracy. The current Russian GLONASS satellite status is accessible via a website.

GLONASS Space Segment
The satellites are in circular orbits with an inclination of 64.8 degrees, and are 19,130 km above the Earth's surface. This means that they are closer to the Earth than are the GPS satellites. A satellite takes 11 h 15 min to orbit the Earth. The position of the constellation repeats every 8 days. GLONASS is undergoing a modernization program whereby new series of satellites are scheduled to be launched. The intention is to expand the constellation to 30 satellites.

GLONASS Navigation Signals
The multiple access technique used by GPS is CDMA, whereas GLONASS uses FDMA. As regards transmissions on the L1 frequency band, GPS uses the carrier frequency 1575.42 MHz whereas GLONASS uses the following frequencies:

$$1602 + 0.5625n$$

where $n = -7, -6, \ldots, 6$

Table 4 Comparison of GLONASS and GPS constellations

	GLONASS	GPS
No. of orbital planes	3	6
Orbital plane inclination (degree)	Approximately 64.2–65.7	55
No. of satellites	24 operational, 0 spares	30 operational
Orbital radius (km)	26,370	26,570

GLONASS uses 14 different frequencies. Each pair of satellites at two antipodal points, that is, points on opposite sides of the Earth that are diametrically opposite one another, is allocated one frequency.

The standard-precision ranging codes are L1OF and L2OF and used on the L1 and L2 frequency bands, respectively. (The codes are analogous to GPS's C/A code.) The obfuscated high-precision ranging codes are L1SF and L2SF and used similarly on the L1 and L2 frequency bands. (The codes are analogous to GPS's P(Y) code.) Table 4 compares the GLONASS constellation with the GPS constellation. Table 5 compares the signal characteristics of GLONASS with those of GPS.

The latest satellite series that is in service is GLONASS-K1. There are plans to launch other satellite series—GLONASS-K2 (from 2018 to 2024) and GLONASS-KM (from 2025 onwards).

COMPASS/BEIDOU AND REGIONAL GPSSs

In 1994 the Chinese government authorized the start of the Beidou satellite navigation system. Beidou is the name given by the Chinese to the Big Dipper constellation. Beidou was initiated by the Chinese military. It provides a regional service.

The Compass Navigation Satellite System (CNSS) uses medium Earth orbit (MEO) satellites and the multiple access technique used is CDMA.

Compass will offer two services: an open service, and a restricted (military) service. The open service is being designed to offer positioning accuracy of 10 m, velocity accuracy of 0.2 m/s, and clock synchronization accuracy of 10 ns. The restricted service is being designed to offer better positioning accuracy, the ability to communicate, and ability to provide

Table 5 Comparison of GLONASS' and GPS' signal characteristics

		GLONASS	GPS
Fundamental clock frequency (MHz)		5.0	10.23
Multiple access technique		FDMA	CDMA
Carrier frequencies (MHz)	L1 band	1598.0625, 1598.625, ..., 1605.375	1575.42
	L2 band	1242.9375, 1243.375, ..., 1248.625	1227.60
PRN	Standard-precision chip rate (Mcps)	0.511	1.023
	High-precision chip rate (Mcps)	5.11	10.23 [L1 P(Y)]; 0.5115 [L2 P(Y)]
	Standard-precision code length (c)	511	1023
	High-precision code length (c)	5.11×10^6	6.19×10^{12} [L1 P(Y)]; 10^{12} [L2 P(Y)]
Navigation message	Frame length (bits) & transmission time (seconds)	1500 & 32 (L1OF & L2OF); 500 (estimate) & 50 (estimate) (L1SF & L2SF)	1500 & 30 (satellite blocks IIA, IIR); 1500 & 60 (satellite blocks IIR-M, IIF, III)
	Superframe length (bits) & transmission time (minutes)	7500 & 2.5 (L1OF & L2OF); 36,000 (estimate) & 12 (estimate) (L1SF & L2SF)	37,500 & 12.5 (satellite blocks IIA, IIR); 37,500 & 25 (satellite blocks IIR-M, IIF, III)

system status information to the user. The frequency bands that are envisaged for CNSS are 1192–1215, 1260–1279, 1559–1561, and 1590–1591 MHz.

It uses an active radiopositioning technique based on time difference of arrival (TDOA) information passed among satellites, a ground station, and user equipment. Beidou-1 is a two-way ranging system. This means that signals are emitted from a control center to the satellites and forwarded to the receiver. A signal is then sent back from the receiver to the control center via the satellites. The control center calculates the receiver's position and sends this information to the receiver.

CHAPTER 7

Solution of an Idealized Problem

The measurement technique used in global positioning system (GPS) is the time difference of arrival of signals. In other words, calculations are based on the time taken for a signal sent by a satellite to arrive at the receiver.

In order for a receiver to receive a signal from a satellite it must have a line of sight (LOS). This means that the path should not be obscured by a solid body. Positioning involves the receiver calculating the range (distance) to each satellite, along the LOS. This is done by decoding the signals transmitted from multiple satellites and looking at the time of arrival of each signal from its respective satellite. If the position of the satellites and the ranges to the satellites were known exactly, then the location of the receiver could be found exactly using three satellites. The receiver would lie on a point of intersection of three spheres centered at the satellites and whose radii were the ranges of the satellites.

Unfortunately, the calculated ranges are in error by a certain amount. The sources of errors in the GPS system include: the ionosphere, multipath, noise, the troposphere, the position of the satellites with respect to the receiver, among others. Multipath is the phenomenon where a satellite's signal may reach the receiver more than once. One path could follow the LOS, while other paths could be due to reflections off buildings, or the ground. There used to be a facility called Selective Availability (SA) that allowed the US Navy to deliberately make signals inaccurate for Standard Positioning Service (SPS) users. The procedure involved in transmitting the encrypted code. When in use, SA dominated the errors in GPS receiver. SA is no longer used. The estimated ranges are called pseudodistances.

The principles of determining the position of a single-frequency GPS receiver are relatively simple. The receiver tracks the satellites using information on their orbits, and so the receiver knows how many satellites are in view, N_{sat} say. The positions and pseudodistances of at least four satellites are needed to estimate the position of a receiver, that is, $N_{sat} \geq 4$. The receiver's position is calculated by triangulation. Consider four satellites $(S_k, k = 1, 2, \ldots, 4)$. The speed of light is c. At time t, the receiver receives

Uncertainties in GPS Positioning
http://dx.doi.org/10.1016/B978-0-12-809594-2.00007-1

data sent from each of the satellites at times t_1, t_2, t_3, and t_4. The position of the satellites when the data was sent was (x_k, y_k, z_k), $k = 1, 2, \ldots, 4$, in an Earth-centered, Earth-fixed coordinate system. The coordinates of the user receiver are (x_0, y_0, z_0). Using the formula "speed = distance/time," we have a system of four equations from which the position of the receiver can be calculated. The pseudodistance δ_k for satellite k is:

$$\delta_k = \sqrt{(x_k - x_0)^2 + (y_k - y_0)^2 + (z_k - z_0)^2} + c \cdot \tau \quad , k = 1, 2, \ldots, 4$$

The problem is that although the satellite clocks are atomic clocks, and for the purpose of this explanation we assume are perfectly synchronized, the receiver's clock is an ordinary clock, which is relatively imprecise. The variable τ is a clock bias. At the instant that the signal arrives at the receiver, we do not know what the time is on the satellite's clock. From the system of equations, we can use an iterative method to find a least-squares estimate of (x_0, y_0, z_0), and the value of τ. Fig. 1 illustrates what is happening. The clock bias is the result of a number of factors: the clock offset (the error in the time given by the receiver clock), the clock drift (the rate at which the clock offset is worsening), the delay in sending the signal from the satellite, the antenna cable delay, and the receiver processing delay. For the purpose of this explanation, we will assume the clock bias to be constant.

The solution of the above system of equations results only in an estimation of the position of a receiver. This is principally due to the errors mentioned earlier. The ionosphere and the troposphere can cause propagation delays in the LOS signal. Multipath causes multiple copies of the signal to arrive at the receiver at different times.

The principles that are used to describe the basics of how GPS works are the same for all operational satellite navigation systems, as well as those under development, whether they are for regional or global usage.

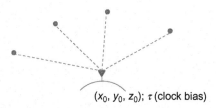

(x_0, y_0, z_0); τ (clock bias)

Fig. 1 Getting a fix on the receiver's position.

Table 1 Pseudodistance calculations ignoring receiver clock offset

Satellite	Transmit Time (s)	Received Time (s)	Difference (ms)	Pseudodistance (km)
1	300	300.063	63	18,900
2	300	300.068	68	20,400
3	300	300.076	76	22,800
4	300	300.095	95	28,500

Each satellite transmits a wave train. This wave train has a pulse, with data being transmitted at each beat. The start of a beat is signified by a particular change in the phase of the signal.

Consider a receiver receiving signals from four satellites. The time that the signal from satellite i arrives at the receiver, according to the receiver's time scale, is $t_{Ri}^u, i = 1, 2, \ldots, 4$. However, the satellites are working in the GPS time scale whereas the receiver is working in its own time scale. The receiver clock offset is the difference between the zero time-points of these two scales, $\tau = t_0^{(u)} - t_0^{(s)}$. Every value of t_{Ri}^u is affected by the receiver clock offset. Table 1 shows example pseudodistances calculated from transmission times given by the satellite and reception times given by the receiver.

From November 2014 to April 2015 the Societe de Calcul Mathematique SA and the Federation Francaise des Jeux Mathematiques jointly ran a competition called the Mathematical Competitive Game 2014–2015 (Beauzamy, 2014). The topic was "Uncertainties in GPS Positioning."

THE COMPETITION

The competition involved five satellites $(S_k, k = 1, 2, \ldots, 5)$ transmitting to a receiver (R). Each satellite transmits a signal and these are processed by the receiver. The difference between the time on the receiver's clock and the time on the satellites' clocks is τ, which is unknown.

The actual distance from the kth satellite to the receiver is d_k (this could also be written as $dist(R, S_k)$), which is unknown. The speed of light is c. The estimate of d_k, calculated by the receiver, is termed the pseudodistance and is denoted by δ_k. This estimate differs from d_k because the time shown on the receiver is invariably different to the time shown on the satellites. The calculation of δ_k is complex as it takes account of various physical

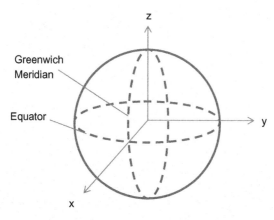

Fig. 2 The axis system.

phenomena—relativity, atmospheric conditions, etc. For the purposes of the competition, these natural effects were ignored.

The reference axes are shown in Fig. 2.

The location of the receiver is (x_0, y_0, z_0). The purpose of the competition is to estimate this location. The position of the kth satellite is (x_k, y_k, z_k). The positions of the satellites are given in Table 2 and the pseudodistances are given in Table 3. All distances are in meters.

Table 2 Satellite positions

	x_k	y_k	z_k
S_1	15470964	−1180726.85	21541839
S_2	19603002	4726671.62	17059949
S_3	3017917	15760014.6	21367886
S_4	180842.8	−15551720.3	21714117
S_5	25616942	7756572.54	336879.3

Table 3
Pseudodistances

	δ_k
S_1	20260438.9
S_2	20264387.99
S_3	23104936.52
S_4	23382913.05
S_5	23101783.81

Table 4 Error distribution of satellite position

Distance from center of sphere (m)	Probability (%)
0–0.4	30
0.4–0.8	25
0.8–1.2	20
1.2–1.6	15
1.6–2	10

The receiver solves the system of five equations:

$$\sqrt{(x_k - x_0)^2 + (y_k - y_0)^2 + (z_k - z_0)^2} + c \cdot \tau = \delta_k$$

to give x_0, y_0, z_0, τ.

Satellites

Table 2, however, only gives approximate figures. The actual position of a satellite is within 2 m of the position shown in Table 2. The error distribution throughout this sphere is nonuniform and is given in Table 4. As can be seen, the actual position of a satellite is more likely to be closer to the approximate position. Within each of the five regions the probability distribution is considered to be uniform.

Pseudodistances

Similarly, Table 3 only gives approximate figures. An actual pseudodistance is within 10 m of the distance shown in Table 3. The error distribution is nonuniform and is given in Table 5. Within each of the 10 regions the probability distribution is considered to be uniform.

Table 5 Error distribution of pseudodistance

Error in approximate pseudodistance	Probability (%)
8–10 m too short	5
6–8 m too short	7.5
4–6 m too short	10
2–4 m too short	12.5
Up to 2 m too short	15
Up to 2 m too long	15
2–4 m too long	12.5
4–6 m too long	10
6–8 m too long	7.5
8–10 m too long	5

Further Assumptions

It is assumed that the satellite position error is independent of its pseudodistance (although, in practice, this is not the case.) Furthermore, it is assumed that both the satellite position error and pseudodistance error of a satellite are independent of the errors in other satellites.

Challenges

Problem 1. What is the expected position of the receiver?

Problem 2. There is a region within which the receiver has a 90% chance of being located. Describe this region.

SOLUTIONS

The competition was directly related to the real-world scenarios that everyday GPS users face, most commonly automobile drivers. For this reason one would expect the end result of a solution to be understandable to a layperson. Even though a mathematical description of the region of interest, in Problem 2, is precise, it is of little practical use to the average GPS user.

SOLUTIONS TO PROBLEM 1

It is possible to use a software tool called a Solver to get an initial approximation of the position of the receiver. An example Solver is one using a least squares approach. One must be aware of the limitation of such an approach. Who is to say that the actual position of the receiver lies at the point where the sum of errors squared is minimized? To find an improved approximation we must take into account the error distributions relating to the satellite positions and pseudodistances. A Solver is unable to do this.

Below is the result of adopting a least squares approach, in layman's terms. This is followed by an explanation of how this approach is undertaken.

SUGGESTED RESULT

The expected position of the receiver is $(4, 343, 409.09; -124, 936.95; 4, 653, 478.56)$. Fig. 3 shows the plan view of the position:

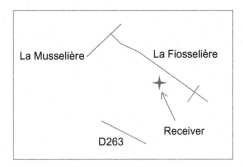

Fig. 3 Plan view of expected position of receiver.

Longitude 1.64764 degrees W of Greenwich.

Latitude 46.96205 degrees N of equator.

The receiver is situated in the Corcoué-sur-Logne, Pays de la Loire, region of France. The nearest city is Nantes lying far to the north. Just to the west of the map is the village of De la Baliniere.

NOTE ON PROBLEM POSED

$S_k = (x_k, y_k, z_k)$: kth "dirty" satellite location, where "dirty" means that it is the raw measurement data.

τ: bias of local clock in the receiver.

When a satellite signal is received and decoded at the receiver, the "dirty" pseudodistance δ_k to the kth satellite can be calculated. The "dirty" geometric distance is d_k, and by taking the local clock bias into account, an estimate of the "dirty" distance is

$$d_k = \delta_k - c\tau.$$

The estimate for the true distance is of course subject to errors. As a consequence we can write

$$d_k + c\tau = \delta_k + e_k.$$

The error term e_k accounts for measurement errors and any inaccuracies in the mathematical model, such as the effect of the ionosphere on c, the velocity of propagation. In the calculation of the expected receiver location, it is necessarily assumed that e_k is an independent random variable with zero mean.

$R = (x_0, y_0, z_0)$: expected receiver location, aka "dirty" receiver location, where "dirty" means that it has been calculated by assuming that e_k is an independent random variable with zero mean. The location is assumed to be stationary.

PROBLEM 1: METHODOLOGY

Variation of Coordinates

Let the expected location and clock bias be $\mathbf{V}_0$. Given the pseudodistances from five satellites, the objective is to estimate the components of the vector $\mathbf{V}_0 = (x_0, y_0, z_0, \tau)$. Let $\delta_k, k = 1 : 5$, be a set of five measurements. Let $f_k(\mathbf{V}_0)$ be the known function which maps the expected location and clock bias $\mathbf{V}_0$ to an accurate kth measurement. Given the vector $\mathbf{V}_0$, the distance function for the kth satellite is defined as

$$f_k(\mathbf{V}_0) = d_k + c\tau = \sqrt{(x_k - x_0)^2 + (y_k - y_0)^2 + (z_k - z_0)^2} + c \cdot \tau.$$

Then with a measurement error e_k we can write $f_k(\mathbf{V}_0) = \delta_k + e_k$. With all five measurements available, and an obvious notation, the results can be expressed in vector form as $\mathbf{F}(\mathbf{V}_0) = \mathbf{R} + \mathbf{E}$. Note that the function $\mathbf{F}$ necessarily involves the locations of all five satellites.

As $\mathbf{V}_0$ is not known, to find a least-squares solution for $\mathbf{V}_0$ we create a guess $\mathbf{V}_G$, which is in error by an unknown vector $\mathbf{V}_D$ from the truth. We use a first-order Taylor expansion of $\mathbf{F}()$ in terms of $\mathbf{V}_D$ in the guess. That is, let $\mathbf{V}_0 = \mathbf{V}_G + \mathbf{V}_D$. Given the guessed location and clock bias $\mathbf{V}_G$ and the measurement vector $\mathbf{R}$, the least-squares objective is to find $\mathbf{V}_D$ such that the sum of the squares of the five measurement errors in the vector $\mathbf{E}$ is a minimum. Since for a given vector $\mathbf{V}$ the function $\mathbf{F}(\mathbf{V})$ is known, and by assumption $\mathbf{V}_D$ is small, then to the first order we can write $\mathbf{V}_0 = \mathbf{F}(\mathbf{V}_G + \mathbf{V}_D)$ which is approximately $\mathbf{F}(\mathbf{V}_G) + \mathbf{A}.\mathbf{V}_D$ where $\mathbf{A} = Grad\mathbf{F}(\mathbf{V}_G)$ is the matrix of derivatives of $\mathbf{F}(\mathbf{V}_G)$ with respect to the vector $\mathbf{V}_G$, evaluated at $\mathbf{V}_G$. This leads to the equations

$$\mathbf{F}(\mathbf{V}_0) = \mathbf{R} + \mathbf{E} = \mathbf{F}(\mathbf{V}_G + \mathbf{V}_D)$$

which is approximately $\mathbf{F}(\mathbf{V}_G) + \mathbf{A}.\mathbf{V}_D$.

The measurement error vector $\mathbf{E}$ can thus be expressed as $\mathbf{E} = \mathbf{A}.\mathbf{V}_D - \mathbf{B}$ where $\mathbf{B} = \mathbf{R} - \mathbf{F}(\mathbf{V}_G)$ is the vector of observed minus computed measurements. The vector $\mathbf{V}_D$, which minimizes the weighted squared error $\epsilon = \mathbf{E}^T \mathbf{\Omega} \mathbf{E}$, where $\mathbf{\Omega}$ is a diagonal matrix of weights, is obtained by differentiating ϵ with respect to the vector $\mathbf{V}_D$ and setting the result to zero.

$$\mathbf{\Omega} = \begin{pmatrix} 1 & 0 & 0 & 0 & 0 \\ 0 & 1 & 0 & 0 & 0 \\ 0 & 0 & 1 & 0 & 0 \\ 0 & 0 & 0 & 1 & 0 \\ 0 & 0 & 0 & 0 & 1 \end{pmatrix}$$

The result for $\mathbf{V}_D$ is given by

$$\mathbf{V}_D = (\mathbf{A}^T \mathbf{\Omega} \mathbf{A})^{-1} \mathbf{A}^T \mathbf{\Omega} \mathbf{B}.$$

The calculated vector $\mathbf{V}_D$ is then added to the original guess $\mathbf{V}_G$ to form a new (improved) guess $\mathbf{V}_G$. The process is then repeated as often as necessary until the calculated update vector $\mathbf{V}_D$ is sufficiently small.

SOLUTION TO PROBLEM 2

This is tackled in Chapter 10.

REFERENCE

Beauzamy, B.   (2014).   *Mathematical competitive game 2014–2015, Société de Calcul Mathématique SA and the Federation Française des Jeux Mathématiques.* Retrieved from http://scmsa.eu/archives/SCM_FFJM_Competitive_Game_2014_2015.pdf (Accessed 26.04.16).

CHAPTER 8

Sources of Inaccuracy

Global positioning system (GPS; and probably most GNSSs) have vulnerabilities.

DELIBERATE INTERFERENCE WITH GPS SIGNALS

The authority controlling GPS could deliberately broadcast erroneous signals from on-world (the ground segment), or off-world via satellites. A rogue element could disrupt a service. The military of different countries could have projects that focus on disrupting GPS. A rogue element could block GPS signals. Jamming techniques are well known and devices are readily available. The blocked signals could be substituted with those of the rogue element (spoofing). Alternatively, the blocked signals could be re-broadcast with some changes, or with time delays (meaconing).

One of the errors that used to exist was selective availability (SA), a denial of accuracy phenomenon. SA involved deliberately degrading signals used in the SPS service. The US government agreed that it would no longer perform SA. It is possible that it could be reintroduced.

DEGRADATION OF PHYSICAL EQUIPMENT

GPS is an extremely complex system comprises numerous items of equipment, each of which has numerous components. No physical component will operate at optimum efficiency forever. A blunder could occur either in the space segment, the ground segment, or the user segment. A receiver error due to either hardware or software failure could cause a positioning error of any size.

OTHER VULNERABILITIES

Forssell (2009) lists other vulnerabilities not mentioned above:
1. A receiver receives very low powered signals which means that a high-powered signal could saturate the receiver.

Uncertainties in GPS Positioning
http://dx.doi.org/10.1016/B978-0-12-809594-2.00008-3

2. Only a small number of frequencies are used (today's GPS makes use of a single one for general use) and the signal structure is known by all.

3. Spectrum competition which could result in congestion, causing radio-frequency interference (RFI).

4. Unintentional interference.

5. RFI interference in the form of harmonics, caused by external sources.

6. Testing at the system level.

7. *Multipath reflections*. These depend on the antenna type and placement of the antenna. Improvements in antenna design have significantly reduced these errors.

8. *Blunders*. A blunder can result in a positioning error of anything up to hundreds of kilometers:

    (a) Blunders caused by human factors such as GPS employees or users lacking the requisite knowledge or training.

    (b) Blunders due to errors in the design, or implementation, of hardware or software. These could occur in any segment be it ground, space, or user.

    A small mistake at the Ground Segment by either a computer or a human can result in a positioning error of up to 1 m whereas a large mistake can result in a positioning error of anything up to hundreds of kilometers. Similarly, a small mistake at the User Segment can result in a positioning error of up to 1 m whereas a large mistake can result in a positioning error of anything up to hundreds of meters.

9. People dismissing potential problems and becoming over-reliant on GPS. To overcome this, a user should be prepared for when GPS experiences problems. In navigation, traditional skills are set to become an important secondary method. A mariner who has relied on electronic navigation at sea and has experienced equipment failure knows only too well the importance of maintaining proficiency in traditional navigational skills.

The US Department of Transport's Volpe National Transportation Systems Center studied how vulnerable a transportation infrastructure reliant on GPS would be. Their report was published in 2001. It preceded the 9/11 incident.

A key vulnerability in the User Segment is that a GPS receiver has to cope with very low powered signals received from the satellite.

INERTIAL NAVIGATION SYSTEM

Some GPS systems are augmented with other systems, one of which is an inertial navigation system (INS). An INS gives the position of a moving object. To make the calculation, integration is performed. This leads to errors.

ACCURACY

Any measurement taken with a technological device is subject to error. Initially, one might consider the errors to be almost random. However, if we had a large quantity of data of known measurements and the corresponding approximate measurements taken using a device, then we could analyze the data to detect patterns. We could also make statements such as "the error is less than 7.5 m, 95% of the time."

In GPS positioning, accuracy is measured as the distance that the estimated position of the receiver is from the true position of the receiver.

Assume that you have tried to estimate, mathematically, the value of an entity several times. The term precision refers to the standard deviation of the estimates. Another way of looking at it is that the term refers to the standard deviation of $|mean - estimate|$ values.

Consider different processes for estimating a receiver's position. Process A involves estimating a receiver's position once. The process is subject to random noise, of large amplitude. Process A has high accuracy and low precision. Process B too involves estimating a receiver's position once. Process B is subject to random noise, of large amplitude. Furthermore, there is a constant bias in estimating a receiver's position. Process B has low accuracy and low precision. Process C involves undertaking process A a number of times and taking the average. Process C has high accuracy and high precision. This is because the effect of the noise is removed. Process D involves undertaking process B a number of times and taking the average. Process D has low accuracy and high precision.

In two dimensions, accuracy of a process is a measure of the closeness of estimates of the receiver's position to the true position, as shown in Fig. 1. Precision of a process considers the distances of estimates from the true position. Precision of a process is a measure of the consistency of these distances, as shown in Fig. 2.

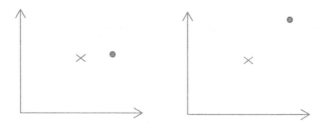

Fig. 1 Accuracy of estimates. The true position is shown as a cross; the estimate is shown as a circle. The estimate on the left is more accurate than that on the right, because it is closer to the true position.

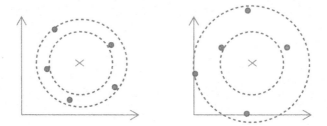

Fig. 2 Precision of estimates. The true position is shown as a cross; an estimate is shown as a circle. The process (for finding estimates) on the left is more precise than that on the right, because the band within which estimates lie is narrower.

In describing how good a GPS is at positioning a receiver, we must only make statements that include both the accuracy and precision components. If one of these components is absent then it leads to ambiguity and may be difficult to interpret. As stated above, we should make statements similar to "the error is less than 7.5 m, 95% of the time."

ERRORS AND ERROR CORRECTIONS

The data used by a receiver to calculate its position, such as a satellite's position, contains errors and so the calculated receiver position is in error by a certain amount.

Receiver positioning accuracy is largely dependent on the accuracy of the pseudo-distances and the satellite ephemeris (ie, satellite orbit data). The user equivalent range error (UERE) is the difference between the true distance from a satellite to the receiver and its calculated pseudo-distance. Receiver positioning accuracy is also affected by how well the satellites are spread out relative to the receiver. The error caused by a weak geometric configuration is referred to as the geometric dilution of position, or simply dilution of precision.

Scope of Errors

Errors are present in the data transmitted from a satellite, further errors can occur during transmission of the data as a signal, yet further errors can occur as the receiver calculates its position.

Communication errors are of little concern. They manifest themselves in the receiver misidentifying bits. However, either forward error correction or cyclic redundancy checking is employed. As a result, a sequence of bits is either accepted (possibly following correction of some bits) or rejected. Hence errors caused by transmission of the signal play no part in the receiver's calculations.

The range of each satellite needs to be calculated as precisely as possible. Consider four satellites whose ranges have been calculated. The ranges are in error by 5, 5, 4, and 9 m, respectively. This could result in the estimated position of the receiver being 20 m, say, from the true position.

Sources of Errors

Errors occur in each of the segments—ground, space, and user. Errors also occur during the transmission of signals. Using this classification, the main sources of error are:

1. Ground segment—ephemeris errors.
2. Space segment—satellite clock bias and satellite code bias.
3. User segment—receiver noise and receiver bias.
4. Signal propagation—ionospheric delay and tropospheric delay (caused by atmospheric effects).

Following is an alternative classification of errors:

1. *Constant errors*: These are errors that have nothing to do with the position of the receiver. The errors are already present in the satellite data before it is propagated. Examples include satellite clock bias and satellite hardware delay.
2. *Correlated errors*: These are errors that are related to the position of the receiver, and where receivers located close to one another have similar error magnitudes. An example group of errors are those due to the medium through which a signal is propagated. This group includes such things as ephemeris errors, ionospheric delay and tropospheric delay. If the magnitude of these errors is known at one receiver location, then it is possible to deduce the magnitude of the relevant errors at a nearby location.

3. *Uncorrelated errors*: These are errors that are related to the position of the receiver, but where one cannot infer the error magnitudes from those of nearby receivers. An example error type is multipath.

Yet another way of classifying GPS errors is specifying the type of error as either noise or bias. One of the bias errors that used to exist was SA.

GROUND SEGMENT

Errors can be introduced in the ground segment.

The monitoring stations track the satellites. This involves calculating the range from the station to each satellite. Data from the monitoring stations is passed to the Master Control Station (MCS). Preprocessing involves making corrections to the measurements that have been passed to it, in order to reduce the effect of errors. These errors are due to the ionosphere, the troposphere, and other propagation effects. The range of a satellite is calculated by a monitoring station using two frequencies. MCS uses the difference between a range measured with one frequency and the same range measured with another frequency to work out a precise measure of the delay in a signal due to the ionosphere. This delay is used to correct the ranges. Meteorological data is also sent from the monitoring stations to MCS. This information is used to make corrections to the measurement errors caused by the signal passing through the troposphere. Models are available to estimate these corrections. During the time between signal transmission and its reception, the earth has rotated, and so a correction needs to be applied to take account of this. The earth has moved with respect to the inertial frame. Corrections also have to be done for relativistic effects. Having made all of the various corrections, the results are smoothed.

A satellite's signal is delayed as it passes through the troposphere. Corrections for the delay are applied at the receiver. However, there is an uncorrected delay of between 0.002 and 0.020 µs. Similarly, a satellite's signal is delayed as it passes through the ionosphere, corrections being applied at the receiver. Once again, there is an uncorrected delay of between 0.002 and 0.020 µs.

Ephemeris Errors

An ephemeris is information from which the position of a satellite can be calculated. The Ground Segment tries to predict the future location of a satellite and errors can occur in these predictions. This is termed ephemeris error.

The positions of satellites are used in the calculation of a receiver's position. Therefore, an ephemeris error will cause an error in positioning the receiver. However, a large ephemeris error is likely to cause only a small receiver positioning error.

An ephemeris data error can result in a positioning error of up to about 1 m.

SPACE SEGMENT

These errors are due to the satellites. Mainly, they include the satellite clock error and the satellite code bias.

Satellite Clock Error

The error in a satellite clock is very small. It differs from true time by a factor of approximately 10^{-13}. If the clock is set to the true time at the start of a day then end of the day, when 86400 s have elapsed, the clock will be in error by about $10^5 \times 10^{-13} = 10^{-8}$ s. Even though the error is small, it accumulates over time. As a ranging calculation involves multiplying by c $(3 \times 10^8 \text{ m/s})$, even small errors can have a great effect on the calculated range. If the clock is in advance by δt_s then the receiver will think that the satellite is closer than it actually is. It will result in a ranging error of

$$b_s = -c\delta t_s$$

The situation is not as bad as it may first appear as the Ground Segment estimates each satellite's clock error. This information is relayed to the respective satellite and forms part of the navigation message propagated to the user receiver. The estimate calculated at the Ground Segment may be slightly in error.

A satellite clock that has not been properly corrected by the Ground Segment can result in a positioning error of up to about 1 m.

USER SEGMENT

Navigation Receiver

A primary source of error in GPS is receiver bias, that is, clock error in the receiver. Satellite geometry can affect the accuracy of GPS positioning. Satellite geometry refers to how dispersed throughout the sky the satellites are that the receiver is using to make its calculations. A complication to the positioning calculation occurs when the receiver is moving.

There are two types of receiver. One determines range using the code phase while the other determines range using the carrier phase. The ranges calculated with code phase-based ranging receivers have larger ranging errors than do carrier phase receivers and, therefore, are less accurate at positioning.

For an uncalibrated receiver, there are hardware delays in the receipt of a satellite's signal by the processing part of the receiver. These are due to the antenna, the antenna cable, and the receiver. The overall delay ranges from 0.005 to 0.250 μs. There is a further delay due to changes in the temperature and environment in which a receiver is operating. This amounts to between 0.001 to 0.005 μs.

There are also setup errors.

SIGNAL PROPAGATION

Data Content

A signal from a satellite contains data, which is used by a receiver for positioning. In addition to the basic data, further data is also sent—satellite clock correction parameters, satellite position correction data, and almanac data for satellites 1 to 24. This data is used to minimize the ranging error.

Many errors depend on the location of the user receiver and so precise corrections cannot be given by the Ground Segment. However, models exist that allow a receiver to estimate corrections.

Data Structure

Data is communicated by satellites to the receiver so that the receiver can calculate its position. The data is digital and is sent using an analog signal.

Each parameter making up the data is represented as a stream of bits. The number of bits used for a particular parameter depends on the range of the parameter and the number of decimal places. For example, assume that there is a parameter that varies between 0 and 1000, and we require accuracy to two decimal places. The possible values of the parameter are $0.00, 0.01, \ldots, 1000.00$. We need, obviously, six decimal digits to represent the parameter's value. Let us consider how many bits we need. There are 100001 possible values. $2^{16} = 65536$; $2^{17} = 131072$. Therefore, 17 bits need to be used. In addition, further bits are added— parity bits. These help the receiver to detect whether or not bits have been corrupted during transmission.

To check whether or not bits have been corrupted during transmission of a signal, checking techniques can be used such as the cyclic redundancy check.

Error Detection and Correction

Channel coding can be applied to digital data before it is transmitted as an analog signal. This means that we do not just transmit the data but we also transmit extra, redundant, bits, the purpose of which is to facilitate error detection. By an appropriate choice of the redundant bits, sometimes erroneous data bits can be not only detected but also corrected. The rate at which data is transmitted, R, may be less than the maximum rate that data can be transmitted through the atmosphere, the channel capacity, C. Shannon's noisy channel coding theorem shows that if $R < C$, then there exists an error-correcting coding technique such that the probability of an error at the receiver is arbitrarily small. The Shannon-Hartley theorem shows that the C is directly related to the bandwidth of the channel.

If $R < C$, this means that we can transmit redundant bits up to the rate $C - R$. The lower R is, the more scope there is for transmitting error detection and correction bits, and the less likely it is for the data to be corrupted by noise. For GPS, R is small. The rate at which the combination of data and redundant bits is transmitted is referred to as the symbol rate. Increasing the symbol rate increases the signal bandwidth.

The extra bits used for checking are referred to as code bits. These bits are used by the receiver to determine whether data bits have been corrupted, and if they have, to try and correct erroneous bits. It may be that the receiver does not need all the bits to be their true values. If the number of erroneous bits is too great then the data cannot be used. The ratio of the number of erroneous bits to the total number of bits is referred to as the bit error rate (BER).

There are two approaches to handling errors in the data received. Firstly, data is repeatedly transmitted from a satellite so if erroneous data is detected then the receiver could simply reject the data and wait for the next occurrence of this data to be sent.

If, every time a bit error occurs, the receiver waits for retransmission then there could be a significant delay. To overcome this we could use the second approach, forward error correction (FEC), to correct some or all of the erroneous bits. If erroneous bits still remain then the receiver could focus on just those bits whose values have yet to be determined. This approach is faster.

There are different categories of error detection and correction methods. One category involves the use of a cyclic code. Another involves the use of a convolution code.

Error Identification

Upon receiving the navigation message, the receiver checks for errors. The syndrome is central to error identification.

Syndrome

The code bits are the extra bits added for error detection and correction. When creating the code bits, the value of each code bit was determined from the original data bits. Knowing how the code bits were generated, the receiver could perform the same operation on the received data bits. If no errors occurred then the sent code bit stream and the received code bit stream would be identical.

Consider the transmission by a satellite of a message $\mathbf{M}$, comprising k bits. Rather than transmitting only the k bits, we transmit extra bits, termed parity bits. The actual sequence of bits that is transmitted is $\mathbf{C}$ and comprises n bits. A matrix $\mathbf{G}$ is used to calculate the values of the parity bits:

$$(\mathbf{M})(\mathbf{G}) = (\mathbf{C})$$

where $\mathbf{M}$ is the original message and is a k-bit row vector; $\mathbf{G}$ is $k \times n$; and $\mathbf{C}$ is a n-bit row vector.

$$\mathbf{G} = \begin{pmatrix} 1 & 0 & 0 & \cdots & 0 & 0or1 & 0or1 & 0or1 \\ 0 & 1 & 0 & \cdots & 0 & 0or1 & 0or1 & 0or1 \\ \vdots & & & & & & & \\ 0 & 0 & \cdots & 0 & 1 & 1or1 & 0or1 & 0or1 \end{pmatrix} = (\mathbf{I}|\mathbf{P})$$

The transmitted bits, $\mathbf{C}$, comprise the original message bits plus extra bits, parity bits. The right hand side of $\mathbf{G}$, that is, $\mathbf{P}$, is chosen to maximize the identification of erroneously transmitted bits. $\mathbf{G}$ determines how each parity bit in $\mathbf{C}$ relates to one or more bits from $\mathbf{M}$.

At the receiver end, let $\mathbf{V}$ be the sequence of bits received. We make use of the matrix $\mathbf{H} = (\mathbf{P}^T|\mathbf{I})$. The value of $\mathbf{S}$, where $\mathbf{S} = \mathbf{V} \cdot \mathbf{H}^T$, determines whether or not any bits have been corrupted during transmission. $\mathbf{S}$ is referred to as the syndrome. If $\mathbf{S} = 0$ then no bits have been corrupted, otherwise corruption has taken place. In the case of corruption, the value of $\mathbf{S}$ is used to help detect and correct corrupted bits.

Example

The original message, $\mathbf{M} = \begin{pmatrix} 0 & 1 & 1 & 0 \end{pmatrix}$. The transmitted sequence, $\mathbf{C} = \begin{pmatrix} 0 & 1 & 1 & 0 & 1 & 1 & 1 \end{pmatrix}$. The received sequence $\mathbf{V} = \begin{pmatrix} 1 & 1 & 0 & 1 & 0 & 0 & 0 \end{pmatrix}$. In this simple example, many of the bits have been corrupted during transmission.

Symbol Rate

Sending one bit at a time from a satellite is less error prone but slower than sending bits in parallel. This is suited to navigation purposes as it is important to strive for error-free reception at the receiver, and the navigation message only comprises a relatively small number of bits.

Error Magnitudes

Noise in the transmission of PRN can result in a positioning error of up to about 1 m. Multipath can result in a delay in a satellite signal reaching the receiver of between 0.002 and 0.010 μs.

RANGING AND POSITIONING

Ranging

Electromagnetic waves do not travel from a satellite to a receiver at constant velocity. To simplify calculations we assume that they do, which leads to errors. The range calculation makes use of this velocity, the time at the receiver when the signal arrived, and the time when the satellite transmitted the signal (this time is part of the data in the signal).

Signals from the satellite are propagated in the radio frequency (RF) band of the electromagnetic spectrum. The atmosphere surrounding the earth causes delays to RF signals. In range calculations, corrections can be made for the effects either by using parameters that are sent with the data or in the case where dual frequencies are used, by comparing two signals. Similarly, corrections can also be made to take account of effects other than those due to propagation.

The clock at the receiver is out-of-sync with the satellite's clocks. There is also ephemeris error. Both of these aspects contribute significantly to ranging and the error terms must feature in the range equation.

An approximate expression for the range is given by the following ranging error model:

$$d_k = \sqrt{(x_k - x_0)^2 + (y_k - y_0)^2 + (z_k - z_0)^2} + c\tau + c\delta t_s + c\delta t_{eph}$$

where the last three terms are receiver clock error, satellite clock error, and ephemeris error, respectively.

Further errors occur during signal propagation and in the user segment. These error terms must also feature in the range equation. The true range, d_k, is calculated from the estimated range by adding the error terms.

$$d_k = \sqrt{(x_k - x_0)^2 + (y_k - y_0)^2 + (z_k - z_0)^2} + c\tau + c\delta t_s + c\delta t_{eph} + c\delta t_{ion}$$
$$+ c\delta t_{trp} + OtherErrors + \epsilon$$

where the errors are comprised of receiver clock error, satellite clock error, satellite ephemeris data error, ionospheric delay to the signal, tropospheric delay to the signal, other errors (such as multipath), and noise. (The signal from a satellite to a receiver does not travel at a uniform velocity but is delayed as it passes through both the ionosphere and the troposphere.)

Due to the errors, the true range is not known exactly. The estimated range is referred to as a pseudo-distance.

The variance of the total pseudo-distance measurement error is the sum of the variances of all independent error sources:

$$\sigma_{PSR}^2 = \sigma_{ReceiverClock}^2 + \sigma_s^2 + \sigma_{eph}^2 + \sigma_{ion}^2 + \sigma_{trp}^2 + \sigma_{OtherErrors}^2$$

Positioning

Having applied corrections to a calculated receiver's position, as much as possible, the position is still in error. In 3D, the true position of the receiver could lie within a sphere centered at the calculated position. Assuming that the most likely position of the receiver is at the center of the sphere, the measure of standard deviation (or variance) can be used to specify the probabilities throughout the sphere.

GPS UERE BUDGETS

The GPS UERE budget components for dual-frequency PPS receivers are shown in Table 1. The figures given for the User Segment are indicative as actual PPS receiver performance varies significantly. Overall, UERE varies

Table 1 Dual-frequency P(Y)-Code UERE budget

Segment	Error source	UERE contribution (m)
Space	Clock stability	0–8.9
	Other space segment errors	1–5.6
Ground	Clock/ephemeris	2.8–9.5
	Other ground segment errors	> 1
User	Ionospheric delay compensation	4.5
	Tropospheric delay compensation	3.9
	Receiver noise and resolution	2.9
	Multipath	2.4
	Other user segment errors	1.0

Table 2 L1 Single-frequency C/A-Code UERE budget

Segment	Error source	UERE contribution (m)
Space	Clock stability	0–8.9
	Other space segment errors	4.1-6.1
Ground	Clock/ephemeris	2.8-9.5
	Other ground segment errors	15.3–25.1
User	Ionospheric delay compensation	Not available
	Tropospheric delay compensation	3.9
	Receiver noise and resolution	2.9
	Multipath	2.4
	Other user segment errors	1.0

between about 7.5 and 13.8 m most of the time. The figures depend on the age of the data in the navigation message and the type of receiver used. The GPS UERE budget components for SPS receivers are shown in Table 2. Once again, the figures given for the User Segment are indicative. Overall, UERE varies between about 12.7 and 24.1 m most of the time.

DIFFERENTIAL POSITIONING

The type of positioning we have described so far is called absolute positioning. The effect of errors on positioning can be reduced by using differential positioning. Here usage is made of a reference receiver. The location of this is known precisely.

We have seen that there are several errors involved in a range calculation. Corrections are applied to reduce the effect of the errors; nevertheless, there is still uncertainty in the estimated position of the receiver.

Differential positioning can remove the errors or further reduce their effect. This is achieved by making use of additional information, that is, the position of the reference receiver.

Correction(s) is(are) passed from the reference receiver to the user receiver. A correction to the user receiver's position could be passed; alternatively corrections to the ranges calculated by the user receiver could be passed.

Overview of Differential Corrections

There are errors involved in calculating the position of a user receiver. However, if we make use of a reference receiver, whose position is known precisely, then a more accurate estimate of the position of the user receiver can be made.

Our calculation of the receiver is dependent on the values of the ranges. At the reference receiver, for a change in a range value we work out the resultant change in the position of the reference receiver.

If the user receiver is close to the reference receiver then the calculated position of the user receiver can be updated by sending correction figures from the reference receiver.

A complication is that the set of satellites seen by the user receiver may be different to the set of satellites seen by the reference receiver. In such circumstances we have to find a common subset, of four or more receivers, that are seen by both receivers. This should be an easy matter as a receiver normally sees about seven satellites.

Error Review
Common Errors (ϵ_0)

These are errors that are common to both the reference receiver and the user receiver. They are independent of the location of a satellite or the user receiver. A common error can vary with time. We can use the function $\epsilon_0(t)$ to represent the error.

An example of a common error is the satellite clock error. If an error is estimated at a particular time in one place then it need not be estimated elsewhere at the same time. With differential positioning, common errors are removed from calculations.

Correlated Errors (ϵ_c)

A correlated error is one where the magnitude of the error at the user receiver can be calculated from the magnitude of the error at the reference receiver. In order to make the calculation we must know the displacement of the user receiver from the reference receiver, along each axis. A correlated error at the user receiver located at $\mathbf{x}$, at time t, is given by

$$\epsilon_c(\mathbf{x}, t) = \epsilon_c(\mathbf{x}_0, t) + \left.\frac{\partial \epsilon_c}{\partial \mathbf{x}}\right|_{\mathbf{x}_0, t} d\mathbf{x} + \frac{1}{2} \left.\frac{\partial^2 \epsilon_c}{\partial \mathbf{x}^2}\right|_{\mathbf{x}_0, t} d\mathbf{x}^2 + \cdots$$

where $\mathbf{x}_0$ is position of the reference receiver and $d\mathbf{x} = \mathbf{x} - \mathbf{x}_0$.

The error $\epsilon_c(\mathbf{x}, t)$ can be calculated exactly if we knew the coefficients of the $d\mathbf{x}, d\mathbf{x}^2, \ldots$ terms. In practice, only the $d\mathbf{x}$ term is used. Hence, only an approximation of $\epsilon_c(\mathbf{x}, t)$ is calculated and the further away $\mathbf{x}$ is from $\mathbf{x}_0$, the worse the approximation.

Errors caused by propagation of a signal are correlated errors. An ephemeris error is also a correlated error. However, the change in ephemeris error is small if the reference and user receivers are in close proximity, and the ephemeris error can be regarded as the same at both locations.

Uncorrelated Errors (ϵ_{uc})

These are errors that are related to the position of the user receiver, but where one cannot infer the error magnitude of the user receiver from the reference receiver. An example error type is multipath.

In summary, for common errors and uncorrelated errors, the magnitude of an error at a user receiver cannot be inferred from the magnitude of the error at the reference receiver. For correlated errors, the magnitude of an error at a user receiver can be inferred from the magnitude of the error at the reference receiver. All of the errors are time dependent, and the time at which an error magnitude applicable to the reference receiver is known precedes the time at which the error magnitude applicable to the user receiver is known. Some types of error change quickly with respect to time while others change slowly. Clock errors are fast-changing, so error magnitudes involving clock errors need to be calculated frequently. Ephemeris errors are slow-changing, so error magnitudes involving ephemeris errors need to be calculated less frequently.

FURTHER READING

Acharya, R. (2014). *Understanding satellite navigation*. Amsterdam: Elsevier.

REFERENCE

Forssell, B. (2009). *The dangers of gps/gnss.* (Coordinates, Feb. http://mycoordinates.org/the-dangers-of-gpsgnss/ Accessed 26.04.16)

CHAPTER 9

Learning From Experience

STUDYING PAST DATA

The signal that leaves a satellite, and before it is distorted, is called the signal-in-space (SIS). Errors originating from both the Ground Segment and the Space Segment are present in SIS. These errors are a major contributor to the inaccuracy of pseudo-distance calculations, and thus to the accuracy of positioning. The errors are called SIS user range errors (SIS UREs). The main causes of SIS UREs are errors in the ephemerides and clock times broadcast by satellites. Researchers have studied this data for broadcasts lasting several years. The usual way to find potential anomalies is to compare the ephemeris and clock data sent with precise data. By doing this they have been able to identify a number of potential SIS anomalies. This approach is called the space approach. However, Heng et al. (2012) have concerns of the ability of the space approach in confirming whether or not a potential anomaly is, in fact, an anomaly. They have undertaken research using an alternative approach in identifying actual anomalies, which will now be described. The approach used was called a top–down approach, aka a ground approach.

Heng et al. (2012) have devised an automatic process for identifying a potential SIS anomaly, and then confirming whether or not it is, in fact, an anomaly. At its starting point, the process has at its disposal the list of potential anomalies identified by the space approach. The process proceeds by looking at the data from ground stations of the International GNNS Service (IGS) and, based on certain requirements, determining preferred IGS stations. The next part of the process involves, for each potential anomaly, taking the pseudo-distance error and subtracting from this nonSIS errors, thus computing the potentially anomalous SIS URE. The final part of the process involves, for each potential anomaly, checking if certain conditions have been met and thus verifying whether or not the potential anomaly is, in fact, an anomaly.

At the time of the research, IGS was a worldwide network of over 350 ground stations that voluntarily track the satellites. An algorithm for ranking the suitability of ground stations was devised. It was decided that, for each potential GPS SIS anomaly, precise data would be collected from between 10 and 32 preferred ground stations of IGS. The automated process in this research downloads receiver independent exchange (RINEX) navigation and observation data for the highest ranked stations (between 10 and 32) and computes SIS UREs for each station.

The pseudo-distance, that is, the measured distance, δ_k is given by:

$$\delta_k = \sqrt{(x_k - x_0)^2 + (y_k - y_0)^2 + (z_k - z_0)^2} + c\tau$$

where c is the speed of light and τ is the difference in time between the receiver's clock and the satellite's clock. The receiver position (x_0, y_0, z_0) is unknown. Both it and τ can be calculated from a system of four or more equations. Let us consider a cruder measure of distance:

$$\rho = \sqrt{(x_k - x_0)^2 + (y_k - y_0)^2 + (z_k - z_0)^2}$$

An expression for ρ is as follows:

$$\rho = d_k - c\tau + I + T + \epsilon'$$

where d_k is the true range, $I = -c\delta t_{ion}$ is the ionospheric delay, $T = -c\delta t_{trp}$ is the tropospheric delay, and ϵ' is a combination of measurement noises, modeling errors, and errors due to unmodeled effects, but excludes SIS URE.

A receiver using the Standard Positioning Service receives the position of a satellite and the time on the satellite clock from the navigation message broadcast by the satellite. The error in the ephemeris contributes an error of ϵ_e in the calculation of the range of the satellite. We have:

$$d_k = d'_k - \epsilon_e$$

where d'_k is the erroneous range estimate that would have resulted from considering the ephemeris data to be true. Similarly, the error in the clock time, after application of the broadcast time correction, contributes an error of ϵ_c in the calculation of the true satellite clock bias. We have:

$$\delta t_s = \delta t'_s - \epsilon_c$$

where $\delta t'_s$ is the erroneous satellite clock bias estimate that would have resulted from considering the satellite clock data to be true. Substituting the expressions for r and δt_s into the equation for the pseudo-distance, and rearranging gives:

$$SIS\ URE = \epsilon_e - c\epsilon_c = \delta'_k + c(\delta t_u - 2\delta t_s - \delta t'_s) + I + T + \epsilon' - \rho$$

where δt_u is the receiver clock bias. (Unfortunately, Heng et al.'s Eqs. (4) and (5) seem to be incorrect.) The values of some of the terms are then estimated.

Knowing a receiver's position and its clock bias would facilitate an estimation of SIS URE, using the above equation. As regards position, the receivers in the IGS network are normally static and so a receiver's position can be found accurately by averaging the many estimates of its position that have been made. After a receiver's position is known and after making estimates of the ionospheric delay, the effects of multipath, the satellite clock bias, and tropospheric delay, we are left with the following expression for the true range of satellite k:

$$d_k = \sqrt{(x_k - x_0)^2 + (y_k - y_0)^2 + (z_k - z_0)^2} + c\delta t_u + \epsilon''_k, \quad k = 1, \ldots, N_{sat}$$

where this time the receiver position (x_0, y_0, z_0) is known and where we ignore satellites that cause potential anomalies.

In this research, a GPS SIS anomaly is defined according to a set of conditions. As stated above, a set of suitable IGS stations were selected (between 10 and 32 of them) and potential anomalies were checked by consulting RINEX observation and navigation data. For each potential SIS anomaly, a decision was made based on how each of the IGS stations categorized the same signal. A potential anomaly became an actual anomaly if at least one of the IGS stations showed anomalous SIS UREs while the remaining stations could not track the satellite during the event.

The automatic verification process was applied to the 31 potential GPS SIS anomalies that were found to have occurred over a period of 8 years, based on the researchers' previous work. Recall that potential anomalies were identified using the space approach. Each SIS anomaly can cause different SIS URE characteristics for different stations. Of the 31 potential anomalies, 26 were found to be true anomalies, one was found not to be an anomaly, and a decision could not be reached on the remaining four as tracking of the relevant satellite was not carried out throughout the anomalous event. For each of the 26 true anomalies, and the false one, the results include the start time of the anomaly, its duration, the maximum URE during the event, and the reference code of the station at which the maximum URE was recorded. The results came from RINEX data at the relevant station. Comparing the results of the ground approach with those of the space approach for the 26 true anomalies, it was found that the start times and durations sometimes differed. The duration of the anomaly given by the ground approach was significantly shorter than that given by the space approach, for almost half of the true anomalies. There were also differences between the two approaches in the values of the maximum UREs recorded.

An example anomaly is the one from the satellite named Space Vehicle Number (SVN) 23 whose signal could be identified from the specific pseudo random noise bit pattern used to generate it, named PRN 29. This occurred on Mar 1, 2007. Another example anomaly was SVN33/PRN03, which took place on July 31, 2006.

For SVN23/PRN 29, the IGS station called "faa1" is where the maximum SIS URE was noted. For SVN33/PRN03, SIS UREs were observed by two stations—"hrm1" and "kuuj." The maximum SIS URE was observed at "htm1" and was 13.14 m. As regards the situation at "kuuj," one of the parameters broadcast in the satellite's navigation message is an upper bound (UB) on the user ranging accuracy (URA). Unfortunately, "kuuj" did not receive the relevant navigation message that updated the URA UB, for reasons unknown, and so recorded SIS UREs. In view of this, the researchers investigated whether any other receiver had not received relevant navigation message on July 31, 2006. They found that several had not.

In conclusion, the research involves the development of an automatic process to carry out verification on potential GPS SIS anomalies. To do this usage is made of data from the IGS network.

MITIGATING RISKS

If a user suspects that the satellite signals are erroneous then he/she should make use of the usual information dissemination channels in order to confirm whether or not there is a known problem. Some applications could combine a receiver with other systems which would enable the application to continue in the event of a GPS system malfunction.

Signal processing techniques could be used to combat malicious interference.

The newer GNSSs (Galileo, INRSS, etc.) are supposedly designed to have better coping strategies with some of the risks, than does GPS.

A user should perform a risk assessment to determine the consequences in the event of GPS problems.

In the United States, there are political and technical reasons why it might not be acceptable to rely on another GNSS as a backup. The technical reason is that all GNSSs operate on similar frequencies and so jamming could easily disable all of them.

LORAN-C was a radio navigation system that had its roots in World War II. It transmitted signal at 100 kHz. In 2010 the US government decided to shut it down. An upgraded version of LORAN-C was developed called eLoran. Due to the difficulties in finding a backup system for GPS, it may be that eLoran will be resurrected to serve that role.

REDUCING ERROR MAGNITUDES WHEN SURVEYING

Ionospheric and tropospheric refraction errors can be reduced by several means. Gathering more redundant data would help. If a base station is being used, then it would be preferable for the baseline length to be short, at most 1km to 5km, say. For longer distances, 20km or more, a dual-frequency receiver should be used. Ignoring observations from satellites whose angle of elevation is lower than, say, 10 degrees to 15 degrees will also reduce the magnitude of these errors.

To reduce the magnitude of multipath errors when performing an RTK survey, the base station receiver should be placed so that it is not near any features that may contribute to these errors.

The geometric dilution of position (GDOP) can be tackled in a number of ways. Tracking many satellites (more than the obligatory four) would help. Satellite geometry is improved when satellite elevations exceed 70 degrees. This should lead to improved position results. Many receivers are capable of tracking between five and twelve satellites simultaneously. GDOP is measured in as a number, the lower the better. The receiver manufacturer may suggest the maximum GDOP number that is acceptable, for example, 7 or 8. A figure of 5 would be ideal.

Setup errors can be reduced by having some way of precisely finding out the receiver antenna's height. In surveying, this is referred to as the antenna reference height (ARH). The equipment in use might have a built-in facility, or it may be available as an accessory. Another alternative is to use fixed-length tripods and bi-pods.

Many of the errors discussed above, including SA (if it ever returns), can be surmounted by using differential positioning. For long baselines

(> 150km), however, sophisticated processing would be required to reduce the effect of errors. The Continuously Operating Reference Station (CORS) was developed by the National Geodetic Survey (NGS) as a nationwide differential measurement system. In 2001, NGS commenced a positioning processing service for dual-frequency, carrier-phase observations. The Geodetic Survey Division (GSD) of Geomatics Canada collaborated with the Geological Survey of Canada to contribute to the Canadian Active Control System. The Canadian products are available upon payment of a subscription. They enable the coordinates of any position in Canada to be found with an accuracy ranging from a centimeter to a few meters.

Prior to carrying out GPS surveys it is important to plan. The almanac data can be studied to establish the time periods when a group of satellites have a good geometry, that is, where each satellite is more than 15 degrees elevation above the horizon. Planning is also used to identify things that may obstruct reception of the signals. Planning software can graphically display GDOP as it changes value throughout the day. The software can also plot several other types of graph to aid the planner. These include the number of satellites available, a polar plot of all visible satellites at a specific time, the satellite orbits as viewed from a specific station on a specific day, and a visibility plot of all satellites over one day. As regards coping with obstructions, a visit to the field should be made to inspect existing stations (ie, places whose coordinates are known) and to put a marker on new stations (ie, places whose coordinates are to be found). For each station, a clinometer could be used to work out, for each obstruction, the angle of elevation that is obscured. For each station, a clinometer (Fig. 1) and a compass could be used to draw a visibility (obstruction) diagram (see Fig. 2), showing the direction and elevation of obstructions and

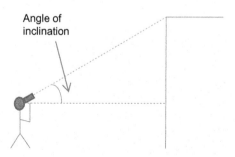

Fig. 1 Using a clinometer.

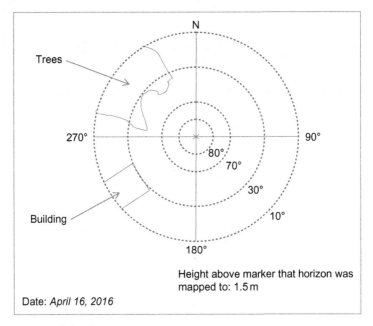

Fig. 2 Station visibility diagram.

potential obstructions. Knowing the satellite orbits the position of any base station can be decided upon. For example, if the satellite orbits are in the southerly sky then the base station, if possible, should be located south of obstructions, so as to avoid interference. If multiple receivers are to be used, then it is best if they are all of the same type, that is, the same number of channels, the same signal-processing software, and the same type of antenna.

REFERENCE

Heng, L., Gao, G., Walter, T., & Enge, P. (2012). Automated verification of potential GPS signal-in-space anomalies using ground observation data. In *2012 IEEE/ION position location and navigation symposium (PLANS)* (pp. 1111–1118). IEEE.

CHAPTER 10

Error Distribution in Data

A question that is seldom posed is: How uncertain are we of the location supplied by a GPS receiver, and how do we calculate this uncertainty?

This chapter continues with the example given in the chapter entitled Solution of an Idealized Problem, which relates to problems posed in a competition (Beauzamy, 2014). This chapter tackles Problem 2. In particular, it takes account of the error distribution of the satellite position (shown in Table 3 of that chapter) and the error distribution of the pseudo-distances (shown in Table 4 of that chapter).

Problem 2. There is a region within which the receiver has a 90% chance of being located. Describe this region.

SOLUTIONS

A Monte Carlo method could be used. We random choose each satellite position, and each pseudodistance, taking account of their respective error distribution. This is then repeated a large number of times. We then feed this data into a Solver to get an estimate of the receiver's position. Unfortunately the Monte Carlo plus Solver approach gives little information about how likely the estimate is of being the true location of the receiver.

Below a superior approach is described (Beauzamy, 2015).

Preparation:

Prep. 1: Subdivide the spheres around each satellite, or radii 2 m, into cubes of side 0.5 m.

Prep. 2: Find an approximate region within which the receiver must lie, either a parallelepiped (3D) or a rectangle (2D).

Prep. 3: Find a minimum and maximum for the time shift. Discretize the interval so that $c\tau$ is approximately 0.5 m; choosing a step size of 10^{-9}s will give a precision of 30 cm. Method:

Uncertainties in GPS Positioning
http://dx.doi.org/10.1016/B978-0-12-809594-2.00010-1

Step 1: Take any position for each satellite (Prep. 1), and position of the receiver (Prep. 2), and any value of the discretized time shift (Prep. 3). Calculate a probability associated with this combination:

prob(this position of satellite S1) × prob(this position of satellite S2) × · · · ×

prob(this position of satellite S5) × prob(this value of the time shift) ×

prob(this value of the pseudodistance to S1) ×

prob(this value of the pseudodistance to S2) × · · · ×

prob(this value of the pseudodistance to S5)

Step 2: Repeat step 1, keeping the receiver position fixed, for all positions of the satellites and all values of the time shift.

Step 3: Repeat steps 1 and 2 for all receiver positions. When done, normalize the list of probabilities so that they sum to 1.

To find the region within which the receiver has a 90% chance of being located, order the list of probabilities from highest to lowest. Next, keep a running total by adding the probabilities from the top until 90% has been reached.

Following is an alternative approach. The result of using it is given first. This is then followed by the method of solution.

SUGGESTED RESULT

Consider a sphere whose center is the center of the Earth and where $(4,343,409.09; -124,936.95; 4,653,478.56)$ is a point on its surface. Now consider the plane that is tangential to this sphere at this point. The true location of the receiver has a 90% chance of lying in the region made up of four layers of rectangular cuboids, two layers lying above the plane (ie, away from the Earth's center) and two layers lying below the plane (ie, towards the Earth's center). The characteristics of each rectangular cuboid are shown in Fig. 1. The make-up of the layers is shown in Figs. 2 and 3.

SUPPLEMENTARY RESULT

Due to the methods devised for Problem 2, it is easy to calculate another result that is useful to those interested in GPS positioning, that is, the region of a map (a 2D entity) within which the receiver has a 90% chance of lying.

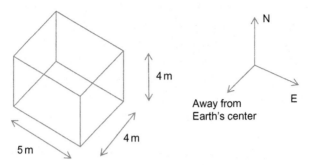

Fig. 1 Dimensions and orientation of rectangular cuboid.

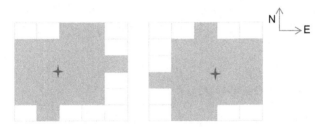

Fig. 2 Plan view of first layer below the plane (left) and first layer above the plane (right). The cross marks the expected position of the receiver. There are 20 rectangular cuboids in each layer.

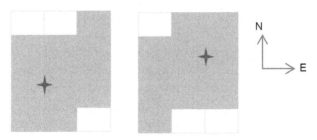

Fig. 3 Plan view of second layer below the plane (left) and second layer above the plane (right). The cross marks the expected position of the receiver. There are 12 rectangular cuboids in each layer.

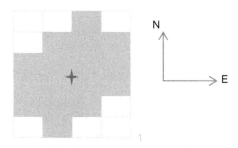

Fig. 4 Region of map within which receiver has a 90% chance of lying. There are 16 rectangles in the region, each with dimensions 5 m (E-W) by 4 m (N-S).

The true location of the receiver has a 90% chance of lying in the region shown in Fig. 4, where the cross is at Longitude 1.64764 degrees W of Greenwich and Latitude 46.96205 degrees N of equator.

PROBLEM 2: METHODOLOGY

R': true receiver location

The approach adopted is to first find an approximate bounding box within which R' must lie. This is followed by subdividing the bounding box into a mesh of rectangular cuboids (small boxes). Finally, the set of small boxes making up the relevant volume is identified, that is, where there is a 90% likelihood that R' lies. This approach involves a number of steps:

a. Deciding on the orientation of the bounding box. The initial idea was to have a bounding box whose faces are parallel to each of the original x-, y-, and z-axes. However, it was realized that a valuable spin-off could be achieved if the axis system was changed such that when taking a plan view of R, a rectangular area on the ground could represent a column of small boxes. Thus the axis system was changed by rotating the original system so that R, from Problem 1, lies on the x-axis (see Note 1 of Appendix E). To simplify the mathematics, it was further decided to translate the axes such that the origin is at R, and the center of the Earth is on the negative x-axis.

b. Finding the $x-$, $y-$, and $z-$extents of the bounding box (see Note 2 of Appendix E).

c. Deciding on the dimensions of a small box and specifying the coordinates of the centroids of all small boxes (see Note 3 of Appendix E).

d. Combining the probability distributions of measurement errors in S_k and d_k into a new distance error distribution (see Note 4 of Appendix E). This involves two sub-steps:

    i. Converting the 3D probability distribution of the measurement error in S_k to a 1D probability distribution (see Note 5 of Appendix E).

    ii. Combining the 1D probability distribution of the measurement error in S_k and the probability distribution of the measurement error in d_k into a new distance error distribution (see Note 6 of Appendix E).

    The result is a piecewise constant function:

    probability $= g$(new distance error)

e. For each satellite, finding a mathematical expression that approximates $dist(P, S_k) - dist(R, S_k)$, where P is any point in the bounding box, and where (x, y, z) are the coordinates of P (see Note 7 of Appendix E).

    From the work here we find that

    $dist(P, S_k) - dist(R, S_k)$ is approximately $b_1 x + b_2 y + b_3 z$.

    As $g()$ is symmetric about zero, this means that the value of g(distance difference) for the centroid at (x, y, z) is the same as it is for the centroid at $(-x, -y, -z)$. This reduces the workload by half!

f. For each small box, calculating the probability that R' lies within the box. Two methods were devised for this step. One is an approximate method (see Note 8 of Appendix E) and the other is a more precise method. For the particular problem that is being solved here, both methods give almost the same solution. The precise method involves several sub-steps:

    i. Working out the dimensions of a mesh of inspection points to be used within a small box (see Note 9 of Appendix E).

    ii. For each satellite, creating a 3D array of increments that describe how g(distance difference) for all of a small box's inspection points can be inferred from the value of g(distance difference) at the small box's centroid (see Note 10 of Appendix E).

    Assume that we have calculated the distance difference for a small box's centroid: $dist(C, S_k) - dist(R, S_k)$. The function $g()$ maps the interval in which the distance difference lies to the probability of it occurring. It is a piecewise constant function. We can find the piece corresponding to the interval in which the distance difference of the centroid lies. The way in which the distance difference varies has been studied in an earlier step. Using this information, we can infer which piece of $g()$ corresponds to each inspection point. We use "-1" to denote the piece directly to the left of the centroid's piece. "−2" denotes the piece that is two pieces to the left of the centroid's piece; "+1" denotes the "piece" directly to the right of the centroid's piece; etc. For example, if $dist(C, S_k) - dist(R, S_k) = 2.2$ the interval in which 2.2 lies is $(2.0, 2.4]$ and so "−1" denotes $(1.6, 2.0]$; "−2" denotes $(1.2, 1.6]$; "+1" denotes $(2.4, 2.8]$.

iii. For each small box, calculating its metric value. The metric that is used, which is directly related to how likely it is for R' to be located inside a small box, is:

$$\Sigma_{AllInspectionPoints}(\Pi_{k=1}^{5} g[dist(P, S_k) - dist(R, S_k)])$$

where $g()$ is the piecewise constant function calculated in an earlier step.

The metric involves first finding g(distance difference) for the centroid (given by $g[dist(C, S_k) - dist(R, S_k)]$) and then inferring $g[dist(P, S_k) - dist(R, S_k)]$ for all of the other inspection points. (This procedure has been mentioned in the previous sub-step.)

We find the metric values for all 512 small boxes ($8 \times 8 \times 8$). The likelihood that a small box contains R', as a percentage, is calculated as follows:

$$\frac{\Sigma_{AllInspectionPoints}(\Pi_{k=1}^{5} g[dist(P, S_k) - dist(R, S_k)])}{\Sigma_{AllSmallBoxes}(\Sigma_{AllInspectionPoints}(\Pi_{k=1}^{5} g[dist(P, S_k) - dist(R, S_k)]))} \times 100$$

Next we create a list of the small boxes sorted by their percentages, from largest to smallest. The region of interest comprises a set of these small boxes. The set is formed by starting at the top of the list and adding each small box to the set until the cumulative probability reaches 90%.

REFERENCES

Beauzamy, B. (2014). *Mathematical competitive game 2014–2015. Socit de calcul mathmatique sa and the federation franaise des jeux mathmatiques.* (http://scmsa.eu/archives/SCM_FFJM_Competitive_Game_2014_2015.pdf/ Accessed 26.04.16)

Beauzamy, B. (2015). *Comments and Results. Mathematical competitive game 2014–2015. Socit de calcul mathmatique sa and the federation franaise des jeux mathmatiques.* (http://scmsa.eu/archives/SCM_FFJM_Competitive_Game_2014_2015_comments.pdf/ Accessed 26.04.16)

CHAPTER 11

Improving Accuracy With GPS Augmentation

Receiver positioning by GNSS is accurate enough for many applications, such as navigation of a car. It is also accurate enough for navigation of ships in open water and for aircraft in flight. However, it is not accurate enough for ships approaching/leaving a harbor or for aircraft landing. Furthermore, positioning by GNSS is unsuited to indoor use due to the attenuation of signals by walls and by the multipath phenomenon. An augmentation system attempts to make corrections for many of the types of error in a GNSS.

The basic GPS constellation is 24 satellites. This could be augmented by using an expanded constellation, such as 30 satellites.

USING TWO GNSSS

Technically, it is not particularly challenging to design and build a receiver that process signals from two GNSSs. GPS could be augmented with another constellation having, say 12 satellites or 24 satellites. Receivers are available that can handle a GPS and GLONASS combination.

Single-Epoch Ambiguity Resolution

The raw data, at the satellite end, is digital yet it is broadcast by an analog signal. Several bits are transmitted in parallel per time period. Each bit pattern corresponds to a unique phase of the analog signal. The receiver continually executes a loop, looking at a part of the signal and deciding what the bit pattern is. The time taken to execute one iteration of the loop is termed an epoch. The process of deciding what the bit pattern is may be difficult as some interference to the signal, or receiver failure, could have occurred. Resolving single-epoch ambiguities has benefits. Effective ambiguity resolution is dependent on the number of signals that a receiver is receiving, the quality of these signals and whether only L1, or both L1 and L2, data is used. The number of signals could be increased by using

Uncertainties in GPS Positioning
http://dx.doi.org/10.1016/B978-0-12-809594-2.00011-3

both GPS and GLONASS. The addition of GLONASS signals complicates matters. By using both the L1 and L2 data it should be possible for a receiver to know its relative position to approximately 10 cm. A number of mathematical methods have been suggested for resolving single-epoch ambiguity resolution, such as the one used by Misra, Pratt, & Burke (1998).

Landing Aircraft

An aircraft can have its own GPS receiver. GPS can be augmented with GLONASS to aid in landing airplanes. The advantage is that availability of the service is significantly improved. The larger the GLONASS constellation, the better the availability. Augmentation is of use when a pilot needs to land an aircraft in poor visibility. GPS with GLONASS can support cockpit instruments. Such a system is only viable if it is available when needed.

A system to aid in landing aircraft comprises aircraft (fitted with receivers), the GPS system, and ground stations to aid the aircraft (fitted with receivers). The system can give both lateral and vertical guidance to the landing instrumentation; the accuracy of vertical guidance is of utmost importance. Service availability of the system comes in two parts— availability within the aircraft, and availability at an airport.

There are other ways of augmenting the basic GPS constellation: using technology at the airport, using geostationary satellites, using more of the GPS satellites than the basic 24 satellites.

Research in using GPS as an aid to landing aircraft has been undertaken by, for example, Pratt, Burke, & Misra (1998).

Augmentation Complications

There are differences between different GNSSs:
1. Different time references
2. Different coordinate frames. For example, two GNSSs could have different Earth-centered, Earth-fixed coordinate frames. This means that the position of satellites and users is expressed differently.
3. Different bandwidths. For example, the chipping rate of the L1 C/A code signal may differ.
4. Different multiple access schemes.
5. Different navigation message content. For example, the number of parameters may differ.

OTHER AUGMENTATION SYSTEMS AND ASSISTED GNSS

To achieve submeter accuracy a GNSS is augmented in one of a number of ways. Further information is provided to help make the calculations more accurate. The extra information is called differential information. One way to do this is using an Earth-based reference station that is located close to a receiver. This is the approach used in Differential GPS (DGPS). DGPS is discussed in Chapter 3. Another approach makes use of additional satellites. Such systems are called satellite-based augmentation systems (SBASs). One advantage of an SBAS is that it improves satellite geometry. This means that the satellites are more spread out across the sky, relative to the receiver. A further approach is called real-time kinematic (RTK) satellite navigation. Yet another approach uses satellite data stored on a server. This approach is called assisted GNSS (AGNSS). With this approach, a mobile phone network could be used to relay from the server to the receiver. Finally, one approach is to use pseudolites (pseudo-satellites). A pseudolite is an Earth-based unit that emits satellite-like signals.

Satellite-Based Augmentation Systems

With an SBAS, usage is made of a satellite(s) that is not part of the constellation. A GPS-compatible signal from an additional satellite can be processed by the receiver. SBAS is also known as a Wide-Area Differential GPS (WADGPS).

Several SBASs exist with others under development. There are commercial and noncommercial SBASs. The existing noncommercial ones are:

1. European Geostationary Navigation Overlay Service (EGNOS), operated by the ESSP (on behalf of EU's GSA). EGNOS began operations in 2003.
2. GPS Aided Geo Augmented Navigation (GAGAN) system being operated by India. GAGAN began operations in 2013.
3. Multi-functional Satellite Augmentation System (MSAS) system, operated by Japan's Ministry of Land, Infrastructure and Transport Japan Civil Aviation Bureau (JCAB). MSAS began operations in 2007.
4. Wide Area Augmentation System (WAAS). It was developed by the US Federal Aviation Administration (FAA) and the Department of Transportation (DOT), and is operated by the FAA. WAAS began operations in 2003. This involves a number of ground stations whose spatial coordinates are known precisely.

5. Wide Area GPS Enhancement (WAGE), operated by the US Department of Defense for use by military and authorized receivers. WAGE began operations in the 1990s.

The noncommercial ones under development are:

1. Quasi–Zenith Satellite System (QZSS), proposed by Japan. QZSS is due to be completed in 2018.
2. GLONASS System for Differential Correction and Monitoring (SDCM), proposed by Russia.
3. Satellite Navigation Augmentation System (SNAS), proposed by China.

Network RTK

RTK is a technique used to enhance positioning by looking at the phase of the carrier wave of the satellite signal rather than PRN. Each carrier signal is modulated by a PRN(s). Consider the L1 frequency band. The period of the carrier wave is $T_{L1} = 1/f_{L1}$ which is approximately 0.635 ns. The time to transmit one chip, T_c, is approximately 1 μs. As a result of the large difference in time between 0.635 ns and 1 μs, RTK is much more accurate than normal GPS ranging. Another way of saying this is in terms of distance. A PRN's chip is comparatively lengthy. The physical distance that each code occupies as it travels from a satellite is shown in Fig. 1. One chip of the C/A code covers a distance of 293 m when transmitted and one chip of the L1 P(Y) code covers 29.3 m. In comparison, the wavelength of L1 is 0.19 m.

With RTK usage is made of a single reference station. The station transmits corrections to mobile receiver, usually as radio waves in the UHF band. At the same time that a satellite is transmitting a signal, comprised of a modulated carrier wave, a receiver is generating the same carrier. The

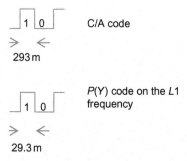

Fig. 1 Physical distance occupied by 1 chip.

receiver must compare the two carrier waves to work out by how much the received wave was delayed during transmission. Unlike chips of the PRN, one period of the carrier wave looks identical to the next. The problem of knowing whether the two waves are matched exactly is called the integer ambiguity problem. The main problem with RTK is in matching the two carrier waves.

Assisted GNSS

Ephemeris data can be sent to a receiver from a server, which is faster than the receiver having to interpret the navigation part of the signal it receives.

With Assisted GNSS (AGNSS) a GNSS receiver is connected to a cellular network so as to improve the receiver's performance. An ordinary receiver, that is, not one connected to a network, takes time to establish the positions of the visible satellites, referred to as the time-to-first-fix (TTFF). It does this by looking at the satellite orbital data (ephemeris) contained in the navigation message. With AGNSS, TTFF is shortened as the ephemeris data is sent to the receiver by a server over the network. Another advantage of GNSS is that the receiver is made more sensitive.

There are two categories of assistance:

1. Mobile Station Based (MSB) where information is used by the receiver (mobile station) to speed up TTFF. The information sent to the GNSS receiver is orbital data (almanac) of the GNSS satellites.
2. Mobile Station Assisted (MSA) where information is sent from the receiver to a server; and the server estimates the receiver position. The server receives good satellite signals and has more computational power than the receiver.

European Geostationary Navigation Overlay Service (EGNOS) is an SBAS. The EGNOS Data Access Service (EDAS) provides access to EGNOS data by ground-based transmission. A server used in AGNSS could benefit from using EDAS.

Assisted GPS Standards

An A-GPS capable device can connect to a server to download information using a network such as CDMA 2000, GSM, LTE, and UMTS. A-GPS protocols are some of the positioning protocols defined by two standardization bodies, 3GPP and Open Mobile Alliance (OMA). The body 3GPP defines protocols for circuit switched networks. The positioning protocols that have been defined are:

1. LPP—for LTE networks;
2. RRC—for UMTS networks;
3. RRLP—for GSM networks;
4. TIA 801—for CDMA 2000 networks.

OMA defines positioning protocols for packet switched networks. Two generations have evolved:

1. SUPL V1.0;
2. SUPL V2.0.

REFERENCES

Misra, P., Pratt, M., & Burke, B. (1998). Augmentation of GPS/LAAS with GLONASS: Performance assessment. In *Proceedings of ION GPS-98*.

Pratt, M., Burke, B., & Misra, P. (1998). Single-epoch integer ambiguity resolution with GPS-GLONASS L1–L2 data. In *Proceedings of ION GPS-98*.

CHAPTER 12

GPS Disciplined Oscillators

In the past, cesium, quartz, and rubidium oscillators were used as frequency standards in calibration laboratories. Increasingly, calibration and metrology laboratories are using a different oscillator as a frequency standard, a Global Positioning System disciplined oscillator (GPSDO). Most GPSDOs in calibration laboratories make use of the C/A ranging code, as broadcast on the L1 frequency band.

HOW A GPSDO WORKS

A GPSDO receives signals from GPS satellites and, based on these signals, controls the frequency of the in-built oscillator. The local oscillator is made from either quartz or rubidium. Disciplined oscillator designers are faced with the problem of how to transfer the time and frequency of a remote oscillator to a local oscillator. Researchers have studied this for several decades and have devised a number of approaches.

A phase locked loop (PLL) is circuitry to which is input a reference signal. Part of the loop includes a voltage controlled oscillator (VCO). The PLL compares the phase of the reference signal to that of the VCO, adjusting the voltage until the two phases match. When this happens we say that the output signal (of the VCO) is locked to the input signal. A part of a GPSDO is a GPS receiver and so the reference input signal to the PLL of a GPSDO comes from the GPS receiver.

With the addition of software to the PLL, the process can be automatically fine-tuned. Compensation can be made for the effects of aging of the local oscillator, temperature, and other environmental parameters. The quality of the substance making up the local oscillator (quartz or rubidium) determines how often compensations need to be made. There is another type of GPSDO design where the local oscillator's frequency is not corrected. In this case, the signal from the oscillator is passed to further circuitry where corrections are made. One can appreciate that GPSDOs are sophisticated instruments that are the result of a vast amount of design effort.

The most difficult part of getting a GPSDO up and running is installing a small antenna on a rooftop. The antenna must have a clear view of the sky. Shapes of antennas differ. For example, it could be conical, where the point of the cone is at the top. Another example shape is a rounded cone. A further shape is similar to a flying saucer (a disc) where the large surface area is uppermost.

A GPSDO has two kinds of output—frequency and time—and these are used as references by laboratories. As regards frequency, this is normally either a 5 MHz or 10 MHz sine wave signal. For the time reference, a 1 pulse per second (pps) signal is output. Time is synchronized to UTC. The back panel of a GPSDO has several sockets for time and several for frequency.

GPSDO PERFORMANCE

Different GPSDO models can perform quite differently. This is especially noticeable if we compare each GPSDO's average performance over a short-time period. Most frequency calibrations undertaken in a calibration laboratory last for a day or less and so in these circumstances it is important that the particular GPSDO being used has an accurate frequency output throughout a one day period. Lombardi (2008) undertook research into the one-day performance of seven different GPSDO models at NIST, which will be described below.

Some models used rubidium local oscillators while other used quartz local oscillators. Rubidium oscillators are much more expensive than quartz ones and are supposed to have technical. However, the rubidium oscillators did always perform better than the quartz ones.

The input signal to a GPSDO uses phase-shift keying. A GPSDO must be capable of quickly locking onto a phase. The research looked at how the local oscillator's frequency varied in with the UTC (NIST) reference frequency. The research looked at the 10-MHz frequency. Further analysis was carried out on two of the GPSDO models—the best performing model and the worst performing model. For each of the two GPSDOs, every hour the average was calculated, the experiment lasting for 80 days. There were large difference between the two GPSDOs—one GPSDOs frequency varied little from 10 MHz while the other GPSDO's frequency differed more markedly. The experiment was repeated using different averaging times—varying from 1 h to about 3 weeks.

The best model had a one-day stability of 6×10^{-14} while the worst model had a one-day stability of 70×10^{-14}. Experiments were also undertaken to measure the short-term stabilities of the two models, with averaging times ranging from 1 to 100 s. With a 100-s averaging time, both models had a stability of 1×10^{-12}. Lombardi suspects that the two models selected probably reflect the extremes of what is available in the marketplace.

Further research has been undertaken by Lombardi, Novick, & Zhang (2005), as regards gauging the performance of GPSDOs. This research will now be described. Four GPSDOs were studied in the research. The researchers had access to four models chosen and they regarded them as being a representative sample of the models commonly used by the governments and industry. The different models had different features. Models A and B each contained an oven controlled crystal (quartz) oscillator (OCXO), while models C and D each a rubidium oscillator. Rubidium oscillators are more expensive and so models C and D were expected to have superior performance, as compared with models A and B.

As regards the configuration of the equipment used, the GPSDOs were all mounted in an equipment rack and all connected to the same antenna, via an antenna splitter. The experiment took place at NIST. The antenna's position was estimated based on precisely known positions in the NIST environment, including geodetic survey markers as well antennas (those used to contribute data to International Atomic Time (TAI)). The coordinates of the antenna were entered into all GPSDOs—latitude, longitude, and altitude. Both the latitude and longitude of models A, C, and D were entered to a certain resolution. However, with model B, a slightly different latitude and longitude was entered due to the model having a different resolution. With some GPSDO models the mask angle can be specified whereas with others it is fixed. Models B and C have fixed mask angles, 5 degrees and 10 degrees, respectively. With models A and D the mask angle can be selected, and both were set to 10 degrees. The antenna cable was an 18.29 m long coaxial cable whose delay had been measured at 73 ns. The antenna splitter had a capability of eight channels; however, only four were used. The combined delay of the antenna cable and antenna splitter was estimated to be 81 ns.

The 1 pps time output and the 10-MHz frequency output were measured simultaneously. One of the reasons for doing this was to see whether the two outputs were in phase with each other. With the four models there were four 1 pps channels. To cope with this a PC-based time measurement system was developed that used a four-port RS-232 interface card connected to four identical time interval counters. (See Novick, Lombardi, Zhang, & Carpentier (1999) for further details of a time interval counter.) The four models also provided four 10 MHz channels and the measurement system used for these was identical to those used by subscribers of the NIST Frequency Measurement and Analysis Service (FMAS). Frequency measurement involved the use of one time interval counter. Readings were taken every 5 s and the 1 h averages were calculated. A test was also conducted on the short-term frequency stability of each GPSDO.

For a GPSDO, there are delays caused by the antenna system, the antenna cable, and the GPSDO itself. The researchers had access to the UTC (NIST) standard and so attempted to measure the delays by comparing the time pulses output from each GPSDO with UTC (NIST), that is, they performed a delay calibration of the GPSDOs. Some GPSDO users do not have access to a reference, such as UTC (NIST), and so the results showed what accuracy could be achieved if a user was able to estimate the delays caused by the antenna system and antenna cable but was unable to estimate the delay caused by the actual GPSDO.

The 1 pps outputs of the four GPSDOs were recorded for 60 days, from March 10th, 2005 to May 8th, 2005. The readings numbered 5,183,926. One minute averages were calculated. None of the GPSDOs seem to lose phase lock with the satellite signals throughout the experiment. Almost all of the readings differed from UTC (NIST) by no more than 50 ns. A difference between the experimental set-up and the actual usage of a GPSDO by a laboratory is that a laboratory does not have a high-precision estimate of the coordinates of the antenna. They have to rely on the GPSDO's ability to survey the antenna's position. The researchers, therefore, decided to repeat the above experiment but this time the antenna's position was not entered at the GPSDO but the GPSDO was left to estimate its position. This was done simultaneously for the four models. For each model, the coordinates of the antenna's position were calculated repeatedly for 10,000 s and the results averaged. Unfortunately, the coordinates obtained by the GPSDOs differed significantly from one another. This was due to differences in the estimation of the antenna's altitude. The latitude and longitude estimates were good. With the self-surveyed antenna position estimated for each model, the experiment described earlier was repeated for 60 days, from May 20th, 2005 to July 18th, 2005. The raw 1 pps data was averaged every 1 min. The results of the two experiments were compared. When using the precise antenna position, the models gave 1 pps outputs in close agreement with UTC (NIST). When using the self-surveyed antenna position, the offsets with UTC (NIST) were greater, and the different models had different magnitudes of offset. During the first experiment, model A had one outlier (ie, large difference with UTC (NIST)); however, during the second experiment there were several outliers. One exceeded 250 ns, while a number exceeded 100 ns. The time stability of each model was calculated for each of the two experiments. Results for averaging times ranging from 1 to 8192 min were derived. The researchers had expected the time stability at short averaging times to be about the same for both cases—known antenna coordinates (KC) and self-surveyed antenna coordinates (SS). They expected this due to the stability of the oscillator. However, they expected that large differences between the KC and SS cases would become apparent once the satellite signals started to discipline the oscillators. The different models behaved differently. In particular, model D, which had the worst short-term stability of the four models, maintained its level of stability even after disciplining kicked in.

Above two experiments have been describe for recording the 1 pps time outputs of four GPSDO models. One experiment used the KC approach and the other used the SS approach. During these experiments, the 10-MHz frequency outputs of the models were also recorded using a system identical to FMAS. The results of two experiments differ little from one another. The large altitude error in the antenna's position calculated using the SS approach caused the 1 pps time outputs to be significantly different from UTC (NIST). However, the altitude error does not significantly affect the frequency in the SS approach. It should be noted that after studying model D, the researchers came to the conclusion that, unusually, the 1 pps output probably did not come from the oscillator. A consequence of this is that the frequency and time outputs for D were not in phase. The long-term frequency stability of the 10-MHz outputs from the four models was analyzed with averaging periods ranging from 1 to 256 h. There were problems with model D for long averaging times.

The short-term frequency stability of each model was calculated by comparing the 10-MHz signal from the model with the 10-MHz signal derived from UTC (NIST). Averaging times ranged from 1 to 100 s.

The research supports the fact that GPSDOs are first class time and frequency standards.

CHOOSING BETWEEN A GPSDO AND A RUBIDIUM OR CESIUM STANDARD

When a calibration laboratory is deciding to buy a primary frequency standard it is likely to choose from a cesium oscillator, a rubidium oscillator, and a GPSDO. In order to facilitate this choice we can look at the performance characteristics. A GPSDO with rubidium in it is superior to a standard rubidium oscillator and so would seem preferable, although some other factors have to be taken into consideration. Deciding between a cesium standard and a GPSDO, however, is more problematic. Cesium oscillators are expensive but this does not mean that GPSDO is the natural choice. There are other criteria and some calibration laboratories will purchase a cesium standard.

Cesium standards are a trusted standard. Unfortunately, there is a tendency for laboratory staff to regard GPSDOs as a trusted standard. It has been known for a GPSDO, and the GPS system it uses, to develop a fault and for laboratory staff to be unaware of this, continuing to use it as a standard. An example failure is the unavailability of GPS signals in a local area. When no signal is available the GPSDO continues to output a frequency signal. Unknown to the user, the GPSDO has gone into cruise mode, which is referred to as holdover. Experiments have been conducted to determine how good this holdover facility is. This was done by measuring the accuracy of the frequency output by a GPSDO. The accuracy of a particular model may be maintained for a few days before it goes awry. It is possible that accuracy of another model will worsen, but not significantly, after a week. In such experiments, lost signals can easily be simulated by disconnecting the antenna.

There are other ways in which a GPSDO, and accompanying GPS system, can fail, such as: damaged antenna cable due to animals or maintenance staff, dislodged antenna due to adverse weather, failure of the local oscillator, jamming. There has even been a case where someone was using an antenna for target practice.

Nevertheless, in common with cesium oscillators, GPSDOs have long-term accuracy and stability. Unlike cesium oscillators, GPSDOs are relatively inexpensive to purchase and to maintain. Furthermore, GPSDOs are continually being adjusted so as to agree with the signals broadcast by GPS satellites, and so are self-calibrating standards. In conclusion, the use of a GPSDO as a primary standard of time and frequency in calibration laboratories has become accepted by most, however, some have their reservations. One aspect that needs to be considered is the determination of a GPSDO's measurement uncertainty. Its traceability needs to be established. There are various ways in which the degree of uncertainty of a GPSDO measurement can be ascertained.

Finally, GPSDOs offer first class performance at a reasonable cost. As a result, they are widely accepted as a primary frequency standard by calibration and testing laboratories.

GPSDOs are indispensable to many applications and certain technologies could not exist without them.

REFERENCES

Lombardi, M. (2008). The use of GPS disciplined oscillators as primary frequency standards for calibration and metrology laboratories. *Measure, 3*(3), 56–65.

Lombardi, M., Novick, A., & Zhang, V. (2005). Characterizing the performance of GPS disciplined oscillators with respect to UTC(NIST). In *Proceedings of the joint IEEE international frequency symposium and precise time and time interval (PTTI) systems and applications.*

Novick, A., Lombardi, M., Zhang, V., & Carpentier, A. (1999). A high performance multi-channel time-interval counter with an integrated GPS receiver. In *Proceedings of the 31st annual precise time and time interval (PTTI) meeting* (pp. 561–568).

Appendix A

Calibration of a GPS Receiver Used as a Time Reference

The US Naval Observatory (USNO) has a master clock facility at Schreiver Air Force Base, near Colorado Springs, Colorado (the same location as GPS' Master Ground Station). The time standard used is a variant of Coordinated Universal Time, called UTC(USNO). GPS satellites broadcast time—GPS time. GPS receivers can be used as a time reference by laboratories. It is possible to calibrate such a receiver so that it provides an on-time 1 pulse per second (pps) reference with a known uncertainty. There are delays to a signal being processed by a receiver that are caused by the antenna and the receiver itself. If a receiver has not been calibrated to account for these delays then the timing pulse output from the receiver is usually within 1 μs of a UTC standard kept at NIST—called UTC(NIST). Lombardi et al. (n.d.) describe a system for remotely calibrating a GPS timing receiver and determining the time offset of the receiver with an uncertainty of <50 ns.

GPS time, UTC(USNO), and UTC(NIST) all differ from one another by small amounts. NIST and USNO have agreed that their times will not differ by more than 100 ns. At the time of the research, the average difference was less than 10 ns. Obviously, GPS time differs little from UTC(USNO), however, a GPS receiver and its antenna can increase the uncertainty of the received time. It is possible to send a receiver and its antenna to NIST, or another laboratory, so that the signal delays that they cause can be estimated. In calibrating, NIST makes use of common-view measurements. Their experience in this technique dates back almost as far as the launch of the first GPS satellite in 1978. Calibration, using common-view measurements, has been undertaken by NIST and other national metrology institutes for many years. Let us explain what this common-view technique is by using an analogy. Imagine that you wish to compare the time on your watch with the time on your friend's watch, when your friend lives at the other side of your town. You both agree on an event, for example, when a certain church's bells begin to chime

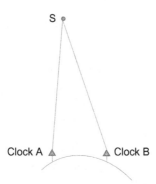

Fig. 1 Use of common-view in GPS

on a Sunday morning. When the event occurs, you each write down the time shown on your watch. You then compare the times. Common-view measurements involving GPS are performed by two receivers, located on Earth, receiving the same signal, see Fig. 1. For each receiver, there is a delay in the satellite's signal reaching the antenna, there is a delay caused by the antenna, and there is a delay caused by the receiver itself. If we perform the calculation (Time on Clock A—Time on Clock B) then the common errors will cancel out. This will leave us with a term $(d_A - d_B)$, where d_A is the antenna and receiver delays caused at A, d_B is the antenna and receiver delays caused at B. Assuming that one of the clocks is accurate, the term $(d_A - d_B)$ can be applied to the other clock as a correction.

In this work, a measurement system was used. This comprised a circuit board that contained a time interval counter (TIC) and a GPS receiver. The procedure is for the measurement system to be sent to the customer, who connects it to a PC. At the customer's site, there is the receiver to be tested and the measurement system (another receiver). The 1-pps output from the receiver under test is compared with the 1-pps output from the measurement system. At NIST, there is an identical measurement system which is compared with a 1-pps output reference signal from UTC(NIST). Control of the process of recording and comparing data was done by software. At both the customer's site and NIST, the software calculated the time difference between pulses every second.

In developing the measurement system, several tests were made. First, a 1-pps signal was sent to two identical measurement systems, both located at NIST. The outputs of the measurement systems were slightly different due to noise. Next a 1-pps signal was sent to two proprietary receivers and the difference in their 1-pps outputs was noted. The difference in outputs was averaged every 10 min and the experiment lasted for approximately 30 days. Several similar comparisons of pairs of receivers were made.

Above, the method for calibrating a customer's GPS receiver was described. This involved a set up at NIST. A further refinement was made to the procedure. This was the use of an Internet connection between NIST and the customer. When developing this approach, the customer selected was Sandia National Laboratories. The distance between the two sites is about 561 km. Data at Sandia was recorded locally and uploaded to NIST once a day, where it was processed. Unfortunately, different antenna cables were used at the two sites, which caused discrepancies. Ideally, NIST should ship an antenna and cable, which has been calibrated, to the customer. The total receiver delay at each site was, therefore, different. (The total receiver delay includes delays caused by the receiver, the antenna, and the antenna cable.) For example, the antenna delay at NIST was estimated to be 408 ns while that at Sandia was estimated to be 161 ns. The total receiver delay at NIST was estimated to be 180 ns, while that at Sandia was estimated to be 427 ns. This remote calibration of Sandia's GPS receiver occurred in the 40-day period, October 8 to November 16, 2002. In conclusion, Lombardi, Novick, & Graham (n.d.) performed a remote calibration of a GPS timing receiver.

REFERENCE

Lombardi, M., Novick, A., & Graham, R. (n.d.). Remote calibration of a GPS timing receiver to UTC(NIST) via the Internet. In *Proceedings of the measurement science conference, Anaheim, CA.*

Appendix B

Comparing Time and Frequency Standards Between Laboratories

Lombardi, Novick, Lopez, Boulanger, & Pelletier (2005) describe the establishment of a comparison network for the Inter-American Metrology System (SIM), which is a body comprising of many national metrology institutes (NMIs). The establishment of this comparison network will now be summarized. Since May 2003, a goal of the regions that made up SIM had been to have some means of ensuring the uniformity of measurements and to strengthen traceability back to SI. It thus created a time and frequency comparison network. This was challenging due to SIM's region being large, the limited resources, and the widely different qualities of standards from country to country. Having said that, at the time, 13 countries were either a member of the BIPM's Metre Convention or an associate of the General Conference on Weights and Measures (CGPM). Of these 13, at least 10 maintained a time and frequency standard, seven of which maintained atomic oscillators and contributed to UTC. The comparison network that was developed used the common-view GPS approach and was capable of making automatic, and continuous, near real-time comparisons. The initial operational network involved the comparison of standards between a Canadian, a Mexican, and a US laboratory. This took place in June 2005. All three countries are members of the region of SIM called North American Metrology Cooperation (NORAMET). This initial network focused on comparing time. The three laboratories were: the National Research Council (NRC), located in Ottawa, Canada; the Centro Nacional de Metrologia (CENAM), located in Queretaro, Mexico; the National Institute of Standards and Technology (NIST), located in Boulder, Colorado, the United States.

The measurement system used will now be described. The common-view technique was used to observe the signal from GPS satellites that had a coarse acquisition (C/A) ranging code. All laboratories used the same equipment. This comprised a computer running Windows 2000 and an LCD monitor (the combination was rack-mounted), and a keyboard which had a trackball. A time interval counter and a GPS timing receiver were mounted onto a circuit board connected to the PC.

The measurement system used an 8-channel GPS timing receiver. This was a similar receiver to those that were being used to make common–view international comparisons. The GPS antenna and its cable were calibrated before they were sent to a laboratory.

In some GPS common–view time and frequency comparisons, usage is made of a schedule of the GPS orbits (a tracking schedule) provided by the BIPM. With the measurement system used by SIM, no tracking schedule was used. A large amount of GPS data was collected and sent via the Internet. To gain some idea of the amount of data, let us look at the situation where a SIM receiver is able to observe eight satellites when they are visible throughout a one day period. In this case, 11,520 min of data (144 segments × 10 min tracks × 8 satellites) would be collected.

NIST undertook the responsibility of supplying the measurement systems to each laboratory. Each system was calibrated prior to shipping. The way in which a system was calibrated was to position two receivers approximately 6 m apart, one was calibrated and the other was being calibrated. The common–view of GPS signals was the approach used. For each receiver, the difference between the 1-pps output and a 1-pps signal conforming to UTC(NIST) was calculated.

As mentioned above, some, a minority, of the NMIs contribute to UTC. The primary way in which comparison of time standards is done is using the common–view technique. Comparison with UTC means that the NMIs standards have traceability back to SI.

Let us turn our attention now to how the data was handled. The SIM measurement system uploaded data every 10 min to an Internet server. The server hosted software to perform data reduction and software to analyze the data. The Web-based analysis software was developed at NIST for a Windows 2000 server. For an individual satellite track, the time difference between two sites could be graphed as 1 h or 1 day averages.

As mentioned above, if the SIM measurement system was able to track eight satellites, when visible, all day then 11,520 min of data would be recorded. Depending upon the relative location of the laboratories, there will be times when only one of the laboratories has line of site with a satellite. The vector that specifies the location of one laboratory with respect to another laboratory is called the baseline. The larger the baseline, the less data is recorded. When a measurement system is being calibrated, the calibrated receiver and the to be calibrated receiver are close to one another and so at any given time, both have line of sight with the same satellites. The SIM region is large and so some baselines are large. At the time of Lombardi et al.'s (2005) paper, these authors were considering amending their data collection and analysis methods so that it took account of all GPS-derived data, regardless of whether or not it was in common view.

The common-view approach requires data to be sent from one laboratory to another and subsequently processed. Hence the results are usually not available in real-time. However, the SIM network gave results in near real-time, the results never being more than 10 min old.

Let us look at the results of the first usage of the comparison network, which involved Canada, Mexico, and the US. NIST's measurement system was operationally ready in April 2005; CENAM's in May 2005; NRC's on June 3rd, 2005. At the time of Lombardi et al.'s paper, data had been collected throughout June, July, and August. Comparisons were made between each pair of laboratories. In addition to this, and independently of this comparison network development, each laboratory used a single-channel GPS timing receiver (ie, it tracks a single satellite) to submit data to BIPM. This data is post-processed by BIPM and this appears weeks after data submission in the Circular-T publication. Comparisons were also made each pair of laboratories using the data in Circular-T. The data in Circular-T is more accurate than that at the laboratories as it makes use of measured ionospheric (MSIO) corrections obtained from measurements taken near a laboratory whereas the comparison network makes use of modeled ionospheric (MDIO) corrections obtained from the satellites' navigation messages. Returning to the SIM comparison network, the mean time offset between each pair of laboratories was calculated. During this initial usage of the network, the mean time offsets in comparisons involving the data submitted by NRC to BIPM were relatively large until a certain day. On that day the altitude of the BIPM receiver's antenna was measured and a correction applied.

Following the initial usage of the comparison network, an uncertainty analysis was carried out. Let us start with the time uncertainty. One type of uncertainty involved studying the standard deviation of the time offset data in each comparison. Another type of uncertainty involves estimating the contribution to time offset of each of a variety of factors. These are shown in Table 1. The antenna position uncertainty is the factor that is likely to vary from one laboratory to another. The overall uncertainty for the time offsets for the CENAM/NIST pair was 15.0 ns, for the NRC/NIST pair it was 15.1 ns, and for the NRC/CENAM pair it was 15.4 ns.

Table 1 Factors contributing to time uncertainty

Factor	Estimated uncertainty (ns)
Antenna position	3
Calibration	5
Operating environment	2
Estimate of "reference delay"	2
Estimate of ionospheric delay	3
Resolution of software	0.5
Time interval counter	2

To estimate frequency uncertainty, a least squares linear line can be fitted to the data to obtain a mean frequency offset.

REFERENCE

Lombardi, M., Novick, A., Lopez, J., Boulanger, J.-S., & Pelletier, R. (2005). The inter-American metrology system (SIM) common-view GPS comparison network. In *Proceedings of the 2005 IEEE international frequency control symposium and exposition* (pp. 691–698). IEEE.

Appendix C

Calibration of GPS Receivers: Comparing Laboratories

Weiss et al. (2011) made a historical comparison of receiver delay calibration between certain labs, covering a period dating back to the 1980s. The rationale for their work was a need to determine the status of GPS receiver delay calibration for some of the receivers whose data contribute to international atomic time (TAI). Different formats for handling GPS data, when comparing clocks, are available. In order to calculate the difference in delays (the differential delay) between a receiver at laboratory A (henceforth referred to as lab A) and a receiver at laboratory B (lab B) usage can be made of a portable system that includes a receiver. The receivers at labs A and B are each, in turn, compared to the portable receiver. We are not only interested in the differential delay between lab A and lab B, but also interested in the uncertainty of the measured delay. A large number of laboratories contribute GPS data towards the generation of TAI.

Since 1980, calibration of a GPS timing receiver has been done by comparing the 1-pps signal output by the receiver with a 1-pps reference signal. Similarly, the frequency output of a receiver can be compared with a frequency reference signal.

Two legacy signals have been broadcast by GPS satellites. One of these has a carrier wave at the L1 frequency and has ranging codes C/A and L1 P(Y). The other has a carrier wave at the L2 frequency and has a ranging code L2 P(Y). The L1 P(Y) and L2 P(Y) ranging codes were not originally intended for civilian use. Several formats exist for storing GPS data and there are several methods for processing it. One of the formats is called RINEX (Receiver Independent Exchange). Data in this format is processed using either software called P3 or software called Precise-Point Positioning.

Weiss et al. (2011) compared the receivers used at the Observatoire de Paris (OP) with those used at NIST, from 1983 onwards. They also compared the receivers used at OP with those used at Physikalisch-Technische Bundesanstalt (PTB), where the receivers processed only the C/A ranging code. This comparison was based on data from BIPM's website. During the period under investigation, OP used two receivers while PTB used three. This change of receivers could account for some of the difference between the time references that were adopted at each laboratory.

Furthermore, Weiss et al. (2011) compared the receivers used at USNO with those used at NIST. Finally, they compared the receivers used at USNO with those used at OP. This comparison was based on data from BIPM's website.

For three laboratories—labs A, B, and C—if we know the differential correction between lab A and lab B, and the differential correction between lab B and lab C, and the differential correction between lab C and lab A, then if we add these corrections the result should be 0. The research looked at such closed loops. In the researchers' work, the mean and the standard deviation of the differential corrections was calculated. The closed loop studies gave results close to 0, taking account of the aggregate of differential correction uncertainties between each pair of laboratories.

If a laboratory calibrates a receiver on its own then it is said to be performing an absolute calibration. The work described above describes finding the difference between receivers at two laboratories. This is referred to as differential calibration (or relative calibration). In generating TAI, only differential calibration is necessary. Over the years, some absolute calibrations have been undertaken at NIST. The first ever primary GPS receiver at NIST was NBS10. A receiver called NIST became the primary one in 2006. In 2007, the US Naval Research Laboratory's (NRL's) receiver and antenna was re-calibrated. This resulted in suggested changes: on the L1 frequency, an advance of 9.4 ns was recommended with an uncertainty of ±0.4 ns; on the L2 frequency, an advance of 2.3 ns was recommended with an uncertainty of ±3.3 ns. The results for the receiver alone were: on the L1 frequency, there was a delay of 28.5 ns with an uncertainty of ±0.4 ns. On the L2 frequency, there was a delay of −45.4 ns with an uncertainty of ±3.3 ns. In April 2011, the USNO's receiver was calibrated. On the L1 frequency, there was a delay of 25.2 ns with an uncertainty of ±0.3 ns. On the L2 frequency, there was a delay of 42.9 ns with an uncertainty of ±0.3 ns. Comparing the L1 frequency results for the 2007 NRL receiver and the 2011 USNO receiver shows an advance of 3.3 with an uncertainty of ±0.5 ns.

In conclusion, the research showed that the calibration results were of mixed quality. This indicates that standard methods for calibration are needed. Regular calibrations are needed, either relative ones or absolute ones.

REFERENCE

Weiss, M., Zhang, V., White, J., Senior, K., Matsakis, D., Mitchell, S., . . . Proia, A. (2011). Coordinating GPS calibrations among NIST, NRL, USNO, PTB, and OP. In *Joint conference of the IEEE international frequency control and the European frequency and time forum (FCS) proceedings* (pp. 1070–1075). IEEE.

Appendix D

Possible Use of a GPS Receiver as an Acceleration Sensor

Research on the possible use of a GPS receiver as an acceleration sensor has been conducted by, for example, Sokolova, Borio, Forssell, & Lachapelle (2010). This will now be described. Consider a receiver on the Earth's surface that is stationary. Any satellite is moving relative to the receiver. Due to the Doppler Effect, the frequency of the carrier signal broadcast from a satellite differs from the frequency of the signal received by the receiver. This difference in frequency is called a frequency shift. For a specific satellite, the frequency shift is not constant; it varies as the satellite moves across the sky. The rate of change of the frequency shift is called the Doppler rate. Sokolova et al.'s research attempts to gauge the acceleration of a moving receiver from Doppler rate measurements. The researchers devised a mathematical model. This was tested using simulated signals and also actual signals. The results seem to confirm that the mathematical model is satisfactory.

REFERENCE

Sokolova, N., Borio, D., Forssell, B., & Lachapelle, G. (2010). Doppler rate measurements in standard and high sensitivity GPS receivers: Theoretical analysis, and comparison. In *Proceedings of the second international conference on indoor positioning and indoor navigation (IPIN)*. Zurich, Switzerland.

Appendix E

Notes on the Error Distribution in Data

The notes that appear in this appendix give further details of a method describe in Chapter 10.

PROBLEM 2: NOTES

1. Rotation matrices:

$$R_x(\theta) = \begin{pmatrix} 1 & 0 & 0, \\ 0 & cos\theta & -sin\theta, \\ 0 & sin\theta & cos\theta \end{pmatrix}$$

$$R_y(\theta) = \begin{pmatrix} cos\theta & 0 & sin\theta, \\ 0 & 1 & 0, \\ -sin\theta & 0 & cos\theta \end{pmatrix}$$

$$R_z(\theta) = \begin{pmatrix} cos\theta & -sin\theta & 0, \\ sin\theta & cos\theta & 0, \\ 0 & 0 & 1 \end{pmatrix}$$

2. Fig. 1 shows the limits of the approximate bounding box.
3. Fig. 2 shows the dimensions of a small box.

   The centroids of all the small boxes are every combination of the following x, y, and z values:

$$x: -14, -10, -6, -2, 2, 6, 10, 14$$
$$y: -17.5, -12.5, -7.5, -2.5, 2.5, 7.5, 12.5, 17.5$$
$$z: -14, -10, -6, -2, 2, 6, 10, 14$$

4. Fig. 3 shows combining of two distributions into one.

5.

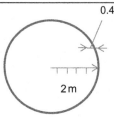

0.4

2 m

Slices of a sphere, each with a thick-ness of 0.4 m.

Volume of sphere $= \frac{4}{3}\pi r^3 = \frac{4}{3}\pi 8$

$$= \frac{32}{3}\pi$$

First case

0.4

0.4

Spherical cap $(r = 2, \ h = 0.4)$

$$\text{Volume} \ = \frac{\pi h^2}{3}(3r - h) = \frac{\pi h^2}{3}(6 - h)$$

$$= \frac{\pi (0.4)^2}{3}5.6$$

Part of the 10% probability region

Volume of 10% probability region = Volume of

whole sphere $-\frac{4}{3}\pi (1.6)^3$

Total probability $= 0.6\%$

Second case

Spherical cap $(r = 1.6, \ h = 0.4)$

Part of the 15% probability region

Shaded volume = Cap $(r = 2, \ h = 0.8)$

$\qquad\qquad\qquad -$Cap $(r = 2, \ h = 0.4)$

$\qquad\qquad\qquad -$ Cap $(r = 1.6, \ h = 0.4)$

Part of the 10% probability region

Total probability $= 2.2\%$

Third case

Spherical cap $(r = 1.2, \ h = 0.4)$

Part of the 20% probability region

Volume adjacent to cap = Cap $(r = 1.6, \ h = 0.8)$

$\qquad\qquad\qquad\qquad\qquad -$Cap $(r = 1.6, \ h = 0.4)$

$\qquad\qquad\qquad\qquad\qquad -$Cap $(r = 1.2, \ h = 0.4)$

Part of the 15% probability region.

Outer volume = Cap $(r = 2, \ h = 1.2)$

$\qquad\qquad\qquad -$Cap $(r = 2, \ h = 0.8)$

$\qquad\qquad\qquad -$Cap $(r = 1.2, \ h = 0.4)$

$\qquad\qquad\qquad -$Volume adjacent to cap

Part of the 10% probability region

Total probability $= 5.3\%$

Fourth case	Spherical cap ($r = 0.8$, $h = 0.4$)

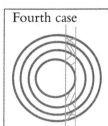	Part of the 25% probability region
	  Volume adjacent to cap $=$ Cap ($r = 1.2$, $h = 0.8$)
	    $-$Cap ($r = 1.2$, $h = 0.4$)
	    $-$Cap ($r = 0.8$, $h = 0.4$)
	Part of the 20% probability region
	  Next volume $=$ Cap ($r = 1.6$, $h = 1.2$)
	    $-$Cap ($r = 1.6$, $h = 0.8$)
	    $-$Cap ($r = 0.8$, $h = 0.4$)
	    $-$Volume adjacent to cap
	Part of the 15% probability region
	  Outer volume $=$ Cap ($r = 2$, $h = 1.6$)
	    $-$Cap ($r = 2$, $h = 1.2$)
	    $-$Cap ($r = 0.8$, $h = 0.4$)
	    $-$Volume adjacent to cap
	    $-$"Next" volume
	Part of the 10% probability region
	Total probability $= 11.6\%$

Fifth case	Hemisphere ($r = 0.4$)

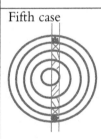	Half of the 30% probability region
	  Volume adjacent to cap $=$ Cap ($r = 0.8$, $h = 0.8$)
	    $-$Cap ($r = 0.8$, $h = 0.4$)
	    $-$Cap ($r = 0.8$, $h = 0.4$)
	Part of the 25% probability region
	  Middle volume $=$ Cap ($r = 1.2$, $h = 1.2$)
	    $-$Cap ($r = 1.2$, $h = 0.8$)
	    $-$Cap ($r = 0.4$, $h = 0.4$)
	    $-$Volume adjacent to cap
	Part of the 20% probability region
	  Next volume $=$ Cap ($r = 1.6$, $h = 1.6$)
	    $-$Cap ($r = 1.6$, $h = 1.2$)
	    $-$Cap ($r = 0.4$, $h = 0.4$)
	    $-$Volume adjacent to cap
	    $-$Middle volume
	Part of the 15% probability region
	  Outer volume $=$ Cap ($r = 2.0$, $h = 2.0$)
	    $-$Cap ($r = 2.0$, $h = 1.6$)
	    $-$Cap ($r = 0.4$, $h = 0.4$)
	    $-$Volume adjacent to cap
	    $-$Middle volume
	    $-$"Next" volume
	Part of the 10% probability region
	Total probability $= 30.2\%$

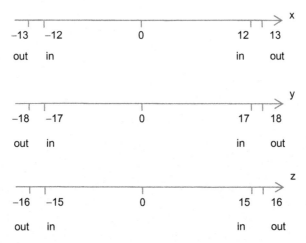

Fig. 1 Delimiting the approximate bounding box.

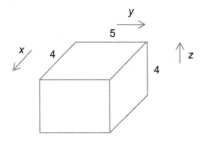

Fig. 2 Dimensions of a small box.

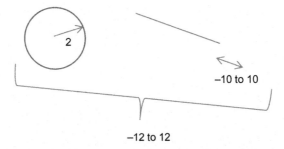

Fig. 3 Combining two probability distributions into one.

6. In what follows we will be looking at points in the vicinity of R. Let us refer to a point under consideration as an inspection point. For a small box, we first handle the centroid C. All other inspection points are considered later.

   S_k is subject to error. From Problem 1, we are able to calculate d_k.

We will use the sign of the error to denote whether the error contributes to C being nearer or further from a satellite as compared to R. For example, a satellite error of between 1.6 and 2 means that we are considering a satellite position that is between 1.6 and 2 m further away from R than S_k than is. A negative satellite error means that we are considering a satellite position that is closer to R than S_k is. Similarly, a distance error of between 9.6 and 10 means that we are considering a distance that is between 9.6 and 10 m longer than d_k. A negative distance error means that we are considering a distance that is shorter than d_k. For example,

a. Distance error = −10 to −8.

b. Satellite error = −2 to −1.6.

Let x% be the likelihood of having this distance error and this satellite error.

The combined error lies between −12 and −9.6. This spans six regions of size 0.4: −12 to −11.6, −11.6 to −11.2, . . . , −10 to −9.6. The probability of the combined error lying in any one of these regions is a fraction of x%. Table 1 shows how the component probabilities are calculated. The columns denote the six regions. The rows denote the distance errors (Table 1).

Table 2 shows the probabilities for different combinations of satellite error (the rows) and distance error (the columns).

Following is a sketch that illustrates the symmetry of the table.

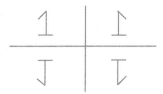

Table 1 Calculating the component probabilities of the six regions

	−12 to −11.6	−11.6 to −11.2	−11.2 to −10.8	−10.8 to −10.4	−10.4 to −10	−10 to −9.6
−8.4 to −8					✓	✓
−8.8 to −8.4				✓	✓	
−9.2 to −8.8			✓	✓		
−9.6 to −9.2		✓	✓			
−10 to −9.6	✓	✓				

Table 2 Probabilities for different error combinations

		1	2	3	4	5	6	7	8	9	10
		≥ -10	> -8	> -6	> -4	> -2	> 0	> 2	> 4	> 6	> 8
1	>1.6	$5a$	$5b$	$5c$	$5d$	$5e$	$5e$	$5d$	$5c$	$5b$	$5a$
2	>1.2	$5f$	$5g$	$5h$	$5i$	$5j$	$5j$	$5i$	$5h$	$5g$	$5f$
3	>0.8	$5k$	$5l$	$5m$	$5n$	$5o$	$5o$	$5n$	$5m$	$5l$	$5k$
4	>0.4	$5p$	$5q$	$5r$	$5s$	$5t$	$5t$	$5s$	$5r$	$5q$	$5p$
5	>0	$5u$	$5v$	$5w$	$5x$	$5y$	$5y$	$5x$	$5w$	$5v$	$5u$
6	> −0.4	$5u$	$5v$	$5w$	$5x$	$5y$	$5y$	$5x$	$5w$	$5v$	$5u$
7	> −0.8	$5p$	$5q$	$5r$	$5s$	$5t$	$5t$	$5s$	$5r$	$5q$	$5p$
8	> −1.2	$5k$	$5l$	$5m$	$5n$	$5o$	$5o$	$5n$	$5m$	$5l$	$5k$
9	> −1.6	$5f$	$5g$	$5h$	$5i$	$5j$	$5j$	$5i$	$5h$	$5g$	$5f$
10	> −2	$5a$	$5b$	$5c$	$5d$	$5e$	$5e$	$5d$	$5c$	$5b$	$5a$

The variable values in the table are:

$$a = \frac{0.6\% \times 5\%}{5}, b = \frac{0.6\% \times 7.5\%}{5}, \ldots, f = \frac{2.2\% \times 5\%}{5}, \ldots,$$

$$\gamma = \frac{30.2\% \times 15\%}{5}.$$

A factor "5" was used to simplify subsequent calculations. It relates to the fact a distance error spans 2 m whereas a satellite error spans 0.4 m.

Tables 3 and 4 show how the combined probability is calculated for each 0.4 m region: -12 to -11.6, -11.6 to -11.2, ..., -0.4 to 0. The left column denotes the row and column numbers of Table 2. It is only necessary to calculate the probabilities in the negative regions as the probabilities are symmetric about 0, that is, the probability of region 0 to 0.4 is the same as that for region -0.4 to 0, the probability for region 0.4 to 0.8 is the same as that for region -0.4 to -0.8, etc. The bottom row of Table 4 is a count of the number of entries in each column. Tables 5 and 6 give further details of how the probabilities are calculated.

7. For a small box, we calculate the distance difference for its centroid (C). For the other inspection points we use an indirect approach (Tables 7 and 8).

    $dist(P, S_1) - dist(R, S_1)$ is approximately $b_1x + b_2y + b_3z$, where P is any point in the bounding box, and where (x, y, z) are the coordinates of P.

    $dist(P, S_1) - dist(R, S_1)$, the distance difference for satellite 1, is in close agreement with a linear model. Tables 9 and 10 show the calculations for satellite 4.

Table 3 Working out the probability that a combined error lies within a specific region of width 0.4 m

	≥ – 12	> – 11.6	> – 11.2	> – 10.8	> – 10.4	> – 10	> – 9.6	> – 9.2	> – 8.8	> – 8.4	> – 8	> – 7.6	> – 7.2	> – 6.8	> – 6.4	> – 6	> – 5.6	> – 5.2	> – 4.8	> – 4.4	> – 4	> – 3.6	> – 3.2	> – 2.8	> – 2.4	> – 2	> – 1.6	> – 1.2	> – 0.8	> – 0.4
10,1	*a*/2	*a*	*a*	*a*	*a*	*a*/2																								
10,2						*b*/2	*b*	*b*	*b*	*b*	*b*/2																			
10,3											*c*/2	*c*	*c*	*c*	*c*	*c*/2														
10,4																*d*/2	*d*	*d*	*d*	*d*	*d*/2									
10,5																					*e*/2	*e*	*e*	*e*	*e*	*e*/2				
9,1		*f*/2	*f*	*f*	*f*	*f*	*f*/2																							
9,2							*g*/2	*g*	*g*	*g*	*g*	*g*/2																		
9,3												*h*/2	*h*	*h*	*h*	*h*	*h*/2													
9,4																	*i*/2	*i*	*i*	*i*	*i*	*i*/2								
9,5																						*j*/2	*j*	*j*	*j*	*j*	*j*/2			
8,			*k*/2	*k*	*k*	*k*	*k*	*k*/2																						
								l/2	*l*	*l*	*l*	*l*	*l*/2																	
													m/2	*m*	*m*	*m*	*m*	*m*/2												
																		n/2	*n*	*n*	*n*	*n*	*n*/2							
																							o/2	*o*	*o*	*o*	*o*	*o*/2		
7,				*p*/2	*p*	*p*	*p*	*p*	*p*/2																					
									q/2	*q*	*q*	*q*	*q*	*q*/2																
														r/2	*r*	*r*	*r*	*r*	*r*/2											
																			s/2	*s*	*s*	*s*	*s*	*s*/2						
																								t/2	*t*	*t*	*t*	*t*	*t*/2	
6,					*u*/2	*u*	*u*	*u*	*u*	*u*/2																				
										v/2	*v*	*v*	*v*	*v*	*v*/2															
															w/2	*w*	*w*	*w*	*w*	*w*/2										
																				x/2	*x*	*x*	*x*	*x*	*x*/2					
																									y/2	*y*	*y*	*y*	*y*	*y*/2

Table 4 Working out the probability that a combined error lies within a specific region of width 0.4 m (continued)

	≥ −12	> −11.6	> −11.2	> −10.8	> −10.4	> −10	> −9.6	> −9.2	> −8.8	> −8.4	> −8	> −7.6	> −7.2	> −6.8	> −6.4	> −6	> −5.6	> −5.2	> −4.8	> −4.4	> −4	> −3.6	> −3.2	> −2.8	> −2.4	> −2	> −1.6	> −1.2	> −0.8	> −0.4
5,1						*u*/2	*u*	*u*	*u*	*u*	*u*/2																			
5,2											*v*/2	*v*	*v*	*v*	*v*	*v*/2														
																w/2	*w*	*w*	*w*	*w*	*w*/2									
																					x/2	*x*	*x*	*x*	*x*	*x*/2				
																										y/2	*y*	*y*	*y*	*y*
4,1							*p*/2	*p*	*p*	*p*	*p*	*p*/2																		
												q/2	*q*	*q*	*q*	*q*	*q*/2													
																	r/2	*r*	*r*	*r*	*r*	*r*/2								
																						s/2	*s*	*s*	*s*	*s*	*s*/2			
																											t/2	*t*	*t*	*t*
3,1								*k*/2	*k*	*k*	*k*	*k*	*k*/2																	
													l/2	*l*	*l*	*l*	*l*	*l*/2												
																		m/2	*m*	*m*	*m*	*m*	*m*/2							
																							n/2	*n*	*n*	*n*	*n*	*n*/2		
																												o/2	*o*	*o*
2,1									*f*/2	*f*	*f*	*f*	*f*	*f*/2																
														g/2	*g*	*g*	*g*	*g*	*g*/2											
																			h/2	*h*	*h*	*h*	*h*	*h*/2						
																								i/2	*i*	*i*	*i*	*i*	*i*/2	
																													j/2	*j*
1,1										*a*/2	*a*	*a*	*a*	*a*	*a*/2															
															b/2	*b*	*b*	*b*	*b*	*b*/2										
																				c/2	*c*	*c*	*c*	*c*	*c*/2					
																									d/2	*d*	*d*	*d*	*d*	*d*/2
																														e/2
10,6																										*e*/2	*e*	*e*	*e*	*e*
9,6																											*j*/2	*j*	*j*	*j*
8,6																												*o*/2	*o*	*o*
7,6																													*t*/2	*t*
6,6																														*y*/2
N	1	2	3	4	5	7	8	9	10	11	12	12	12	12	12	12	12	12	12	12	12	12	12	12	12	12	12	12	12	12

Table 5 Probabilities for each region spanning 0.4 m

≥ -0.4	$y/2 + y + t + o + j + d/2 + e/2 + e + j + o + t + y/2$
≥ -0.8	$t/2 + y + y + t + o + i/2 + j/2 + d + e + j + o + t/2$
≥ -1.2	$o/2 + t + y + y + t + n/2 + o/2 + i + d + e + j + o/2$
≥ -1.6	$j/2 + o + t + y + y + s/2 + t/2 + n + i + d + e + j/2$
≥ -2	$e/2 + j + o + t + y + x/2 + y/2 + s + n + i + d + e/2$
≥ -2.4	$e + j + o + t + x/2 + y/2 + x + s + n + i + c/2 + d/2$
≥ -2.8	$e + j + o + s/2 + t/2 + x + x + s + n + h/2 + i/2 + c$
≥ -3.2	$e + j + n/2 + o/2 + s + x + x + s + m/2 + n/2 + h + c$
≥ -3.6	$e + i/2 + j/2 + n + s + x + x + r/2 + s/2 + m + h + c$
≥ -4	$d/2 + e/2 + i + n + s + x + w/2 + x/2 + r + m + h + c$
≥ -4.4	$d + i + n + s + w/2 + x/2 + w + r + m + h + b/2 + c/2$
≥ -4.8	$d + i + n + r/2 + s/2 + w + w + r + m + g/2 + h/2 + b$
≥ -5.2	$d + i + m/2 + n/2 + r + w + w + r + l/2 + m/2 + g + b$
≥ -5.6	$d + h/2 + i/2 + m + r + w + w + q/2 + r/2 + l + g + b$
≥ -6	$c/2 + d/2 + h + m + r + w + v/2 + w/2 + q + l + g + b$
≥ -6.4	$c + h + m + r + v/2 + w/2 + v + q + l + g + a/2 + b/2$
≥ -6.8	$c + h + m + q/2 + r/2 + v + v + q + l + f/2 + g/2 + a$
≥ -7.2	$c + h + l/2 + m/2 + q + v + v + q + k/2 + l/2 + f + a$
≥ -7.6	$c + g/2 + h/2 + l + q + v + v + p/2 + q/2 + k + f + a$
≥ -8	$b/2 + c/2 + g + l + q + v + u/2 + v/2 + p + k + f + a$
≥ -8.4	$b + g + l + q + u/2 + v/2 + u + p + k + f + a/2$
≥ -8.8	$b + g + l + p/2 + q/2 + u + u + p + k + f/2$
≥ -9.2	$b + g + k/2 + l/2 + p + u + u + p + k/2$
≥ -9.6	$b + f/2 + g/2 + k + p + u + u + p/2$
≥ -10	$a/2 + b/2 + f + k + p + u + u/2$
≥ -10.4	$a + f + k + p + u/2$
≥ -10.8	$a + f + k + p/2$
≥ -11.2	$a + f + k/2$
≥ -11.6	$a + f/2$
≥ -12	$a/2$

Consider three points equally spaced along a line. Fig. 4 shows the distances from a fourth point to each of the three points. The distances are d, e, and f.

$$d = \sqrt{a^2 + b^2}$$

$$e = \sqrt{(a + c)^2 + b^2}$$

$$f = \sqrt{(a + 2c)^2 + b^2}$$

$$e - d = \sqrt{(a + c)^2 + b^2} - \sqrt{a^2 + b^2}$$

Table 6 Probabilities for each region spanning 0.4 m: contribution made by each algebraic variable

Level	a	b	c	d	e	f	g	h	i	j	k	l	m	n	o	p	q	r	s	t	u	v	w	x	y
		1/2	1/2	1/2	1 1/2				1/2	2				1/2	2				1/2	2				1/2	2
−2	1/2	1	1	1	1			1/2	1	1 1/2	1/2		1/2	1	2			1/2	1	2			1/2	1 1/2	2
	1	1	1	1	1	1/2	1/2	1	1	1	1	1/2	1	1	1 1/2		1/2	1	1	2		1/2	1 1/2	2	2
−4	1	1	1	1	1	1	1	1	1	1	1	1	1	1	1	1/2	1	1	1 1/2	1 1/2	1/2	1 1/2	2	2	2
	1	1	1	1	1	1	1	1	1	1	1	1	1	1	1	1	1	1 1/2	2	1	1 1/2	2	2	2	1 1/2
−6	1	1/2	1/2	1/2	1	1	1	1/2	1/2	1	1	1	1	1	1	1	1 1/2	2	1 1/2	1	2	2	2	1 1/2	1/2
	1/2	1/2	1/2	1/2	1	1/2	1/2	1/2	1/2	1	1	1	1	1	1	1 1/2	2	1 1/2	1	1/2	2	2	1 1/2	1/2	
−8	1/2	1	1	1	1	1/2	1/2	1	1	1/2	1	1	1	1	1/2	2	1 1/2	1	1		2	1 1/2	1/2		
	1	1	1	1	1	1	1	1	1		1	1	1	1		1 1/2	1	1	1/2		1 1/2	1/2			
−10	1	1	1	1		1	1	1	1		1	1	1	1/2		1	1	1/2			1/2				
	1	1	1	1		1	1	1/2	1		1	1	1/2			1	1/2								
	1					1	1/2		1/2		1	1/2				1/2									
> −12 1/2	1/2					1/2					1/2														

Table 7 $dist(C, S_1) - dist(R, S_1)$ for centroids at $x = -10$, to 1 d.p

	y = −12.5	−7.5	−2.5	2.5	7.5	12.5
z = 14	...	...	...	...	...	8.0
10	...	...	...	8.3	8.5	8.6
6	...	...	8.8	8.9	9.1	9.3
2	...	9.2	9.4	9.6	9.8	10.0
−2	...	9.9	10.1	10.3	10.5	...
−6	...	...	10.8	10.9	...	...
−10	...	...	11.4	...	...	...

Table 8 $dist(C, S_1) - dist(R, S_1)$ for centroids at $x = -6$, to 1 d.p

	y = −12.5	−7.5	−2.5	2.5	7.5	12.5
z = 14	...	...	...	...	3.9	...
10	...	...	4.2	4.3	4.5	4.7
6	...	4.6	4.8	5.0	5.2	5.4
2	5.1	5.3	5.5	5.7	5.9	6.0
−2	...	6.0	6.2	6.3	6.5	...
−6	...	6.6	6.8	7.0	7.2	...
−10	...	...	7.5	7.7	...	...

Table 9 $dist(C, S_4) - dist(R, S_4)$ for centroids at $x = -10$, to 1 d.p

	y = −12.5	−7.5	−2.5	2.5	7.5	12.5
z = 14	...	...	...	...	...	4.0
10	...	...	...	−0.2	3.1	6.4
6	...	...	−1.1	2.2	5.5	8.9
2	...	−2	1.4	4.7	8.0	11.3
−2	...	0.5	3.8	7.1	10.5	...
−6	...	...	6.3	9.6	...	...
−10	...	...	8.7	...	...	...

Table 10 $dist(C, S_4) - dist(R, S_4)$ for centroids at $x = -6$, to 1 d.p

	y = −12.5	−7.5	−2.5	2.5	7.5	12.5
z = 14	...	...	...	...	−1.1	...
10	...	...	−5.3	−1.9	1.4	4.7
6	...	−6.1	−2.8	0.5	3.8	7.2
2	−7.0	−3.7	−0.3	3.0	6.3	9.6
−2	...	−1.2	2.1	5.4	8.8	...
−6	...	1.2	4.9	7.9	11.2	...
−10	...	...	7.0	10.4	...	...

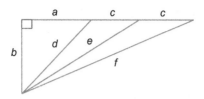

Fig. 4 Example to show that distance differences do not vary linearly.

$$f - e = \sqrt{(a + 2c)^2 + b^2} - \sqrt{(a + c)^2 + b^2}$$

For $dist(P, S_k) - dist(R, S_k)$ to vary linearly, $e - d$ should be equal to $f - e$. However, in general $e - d \neq f - e$ and so a linear model is only an approximation. However, $dist(P, S_k) - dist(R, S_k)$ only differs from $b_1 x + b_2 y + b_3 z$ by of the order of less than 1 mm.

a. $dist(P, S_1) - dist(R, S_1)$ is approximately $-0.98536x + 0.036312y - 0.16666z$.

b. $dist(P, S_2) - dist(R, S_2)$ is approximately $-0.95688x - 0.26107y + 0.127343z$.

c. $dist(P, S_3) - dist(R, S_3)$ is approximately $-0.47628x - 0.68582y - 0.55027z$.

d. $dist(P, S_4) - dist(R, S_4)$ is approximately $-0.42493x + 0.664824y - 0.61436z$.

e. $dist(P, S_5) - dist(R, S_5)$ is approximately $-0.48512x - 0.36763y + 0.793415z$.

8. The inspection points that we use are the centroids (Cs) of the small boxes. The metric that is used, which is directly related to how likely it is for R' to be located inside a small box, is:

$$\prod_{k=1}^{5} g[dist(C, S_k) - dist(R, S_k)]$$

where $g()$ is the piecewise constant function calculated in an earlier step. The metric involves first finding the distance difference for a centroid (given by $dist(C, S_k) - dist(R, S_k)$). Table 11 shows example

Table 11 Distance differences for centroids of two small boxes

	(−2,−2.5, −2)	(−2, −2.5, 2)
S_1	2.2 m	1.5
S_2	2.3	2.82
S_3	3.8	1.57
S_4	0.42	2.04
S_5	0.3	3.5

distance differences for two small box centroids. The figures are correct to 1 d.p. However, where the approximation is a multiple of 0.4, the figures are given to a higher precision. This is so that when we use $g()$, we can avoid a distance difference that falls between two pieces of the piecewise constant probability function (Table 11).

We find the metric values for all 512 small boxes ($8 \times 8 \times 8$). The likelihood that a small box contains R', as a percentage, is calculated as follows:

$$\frac{\Pi_{k=1}^{5} g[dist(C, S_k) - dist(R, S_k)]}{\Sigma_{AllCentroids} (\Pi_{k=1}^{5} g[dist(C, S_k) - dist(R, S_k)]} \times 100$$

Next we create a list of the small boxes sorted by their percentages, from largest to smallest. The region of interest comprises a set of these small boxes. The set is formed by starting at the top of the list and adding each small box to the set until the cumulative probability reaches 90%.

9. Let us see how the distance differences change as we move from one side of a small box to the other.

   a. For satellite 1: Consider $-0.98536x + 0.036312y - 0.16666z$

      i. A movement in the x-direction of 4 m, keeping y and z constant, causes a change of approximately 3.9 in the distance difference.

      ii. Moving 5 m in y-direction only, causes a change of approximately 0.2.

      iii. Moving 4 m in z-direction only, causes a change of approximately 0.7.

   b. For satellite 2: Consider $-0.95688x - 0.26107y + 0.127343z$.

      i. Moving 4 m in x-direction only, causes a change of approximately 3.8.

      ii. Moving 5 m in y-direction only, causes a change of approximately 1.3.

      iii. Moving 4 m in z-direction only, causes a change of approximately 0.5.

   c. For satellite 3: Consider $-0.47628x - 0.68582y - 0.55027z$.

      i. Moving 4 m in x-direction only, causes a change of approximately 1.9.

      ii. Moving 5 m in y-direction only, causes a change of approximately 3.4.

      iii. Moving 4 m in z-direction only, causes a change of approximately 2.2.

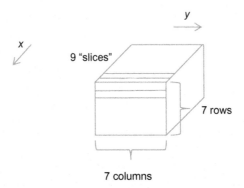

Fig. 5 Mesh of inspection points within a small box.

d. For satellite 4: Consider $-0.42493x + 0.664824y - 0.61436z$.

   i. Moving 4 m in x-direction only, causes a change of approximately 1.7.

   ii. Moving 5 m in y-direction only, causes a change of approximately 3.3.

   iii. Moving 4 m in z-direction only, causes a change of approximately 2.4.

e. For satellite 5: Consider $-0.48512x - 0.36763y + 0.793415z$.

   i. Moving 4 m in x-direction only, causes a change of approximately 1.9.

   ii. Moving 5 m in y-direction only, causes a change of approximately 1.8.

   iii. Moving 4 m in z-direction only, causes a change of approximately 3.2.

Based on the above analysis, and the fact that the piecewise constant function, $g()$, has steps of 0.4m wide, we use a mesh $9 \times 7 \times 7$ inspection points in the x, y, and z directions respectively, as shown in Fig. 5. C is the center inspection point. The rows of inspection points are 0.4m apart in each of the x, y, and z directions (Fig. 5).

10. $c = g[dist(C, S_k) - dist(R, S_k)]$

Increments: "-1" denotes the probability is that of the piece of $g()$ to the left of where c has been derived from.

Slices: "$+1$ slice" is the slice one away from the center slice in the positive x-direction (Tables 12–17).

Table 12 Satellite 1 increments

Slice	Increment for all inspection points in slice
Middle	0 (ie, probability is c)
+1	−1
+2	−2
+3	−3
+4	−4
−1	+1
−2	+2
−3	+3
−4	+4

Table 13 Satellite 2 increments

Slice	Smallest y						Largest y
Middle	1	1	0	0	0	−1	−1
+1	0	0	−1	−1	−1	−2	−2
+2	−1	−1	−2	−2	−2	−3	−3
+3	−2	−2	−3	−3	−3	−4	−4
+4	−3	−3	−4	−4	−4	−5	−5
−1	2	2	1	1	1	0	0
−2	3	3	2	2	2	1	1
−3	4	4	3	3	3	2	2
−4	5	5	4	4	4	3	3

Increment. All rows in a slice are the same

Table 14 Satellite 3 increments for middle slice

Increment

	Smallest y						Largest y
Largest z	1	0	−1	−2	−3	−4	−5
	2	1	0	−1	−2	−3	−4
	2	1	0	−1	−2	−3	−4
	3	2	1	0	−1	−2	−3
	4	3	2	1	0	−1	−2
	4	3	2	1	0	−1	−2
Smallest z	5	4	3	2	1	0	−1

Table 15 Satellite 4 increments for middle slice

	Smallest y						Largest y
				Increment			
Largest z	−5	−4	−3	−2	−1	0	1
	−4	−3	−2	−1	0	1	2
	−4	−3	−2	−1	0	1	2
	−3	−2	−1	0	1	2	3
	−2	−1	0	1	2	3	4
	−2	−1	0	1	2	3	4
Smallest z	−1	0	1	2	3	4	5

Table 16 Satellite 5 increments for middle slice

	Smallest y						Largest y
				Increment			
Largest z	3	3	2	2	2	1	1
	3	3	2	2	2	1	1
	2	2	1	1	1	0	0
	1	1	0	0	0	−1	−1
	0	0	−1	−1	−1	−2	−2
	−1	−1	−2	−2	−2	−3	−3
Smallest z	−1	−1	−2	−2	−2	−3	−3

Table 17 Satellites 3, 4, 5 increments for nonmiddle slices

Slice	Increments
+1	same as middle slice
+2	as middle slice with 1 subtracted from all
+3	same as +2 slice
+4	same as +2 slice
−1	same as middle slice
−2	as middle slice with 1 added to all
−3	same as −2 slice
−4	same as −2 slice

BIBLIOGRAPHY

Acharya, R. (2014). *Understanding satellite navigation*. Amsterdam: Elsevier.

Ashby, N. (2003). Relativity in the global positioning system. *Living Reviews in Relativity*, *6*(1), Retrieved from http://relativity.livingreviews.org/Articles/lrr-2003-1/ (Accessed 26.04.16).

Beauzamy, B. (2014). *Mathematical competitive game 2014–2015, Société de Calcul Mathématique SA and the Federation Française des Jeux Mathématiques*. Retrieved from http://scmsa. eu/archives/SCM_FFJM_Competitive_Game_2014_2015.pdf (Accessed 26.04.16).

Beauzamy, B. (2015). *Comments and results, mathematical competitive game 2014–2015, Société de Calcul Mathématique SA and the Federation Française des Jeux Mathématiques*. Retrieved from http://scmsa.eu/archives/SCM_FFJM_Competitive_Game_2014_2015_comments.pdf (Accessed 26.04.16).

Calcutt, D., & Tetley, L. (1994). *Satellite communications: Principles and applications*. Oxford: Butterworth-Heinemann.

Cooksey, D. (n.d.). *NAVSTAR global positioning system (GPS) facts*. Retrieved from http://www.montana.edu/gps/NAVSTAR.html (Accessed 26.04.16).

Dardari, D., Luise, M., & Falletti, E. (Eds.) (2011). *Satellite and terrestrial radio positioning techniques: A signal processing perspective*. Amsterdam: Elsevier.

Forssell, B. (2009). The dangers of GPS/GNSS. *Coordinates*, Retrieved from http://mycoordinates.org/the-dangers-of-gpsgnss/ (Accessed 26.04.16).

Heng, L., Gao, G., Walter, T., & Enge, P. (2012). Automated verification of potential GPS signal-in-space anomalies using ground observation data. In *2012 IEEE/ION position location and navigation symposium (PLANS)* (pp. 1111–1118). IEEE.

Jung, J., Enge, P., & Pervan, B. (n.d.). *Optimization of cascade integer resolution with three civil GPS frequencies*. Retrieved from http://gps.stanford.edu/papers/Jung_IONGPS_2000.pdf (Accessed 26.04.16).

Kavanagh, B., & Slattery, D. (2014). *Surveying with construction applications* (8th ed.). Pearson: London, New York.

Lombardi, M. (2008). The use of GPS disciplined oscillators as primary frequency standards for calibration and metrology laboratories. *Measure*, *3*(3), 56–65.

Lombardi, M. (2012). Microsecond accuracy at multiple sites: Is it possible without GPS? *IEEE Instrumentation & Measurement Magazine*, *2012*, 14–21.

Lombardi, M., Nelson, L., Novick, A., & Zhang, V. (2001). Time and frequency measurements using the global positioning system. *Cal Lab: The International Journal of Metrology*, *8*(3), 26–33.

Lombardi, M., Novick, A., & Graham, R. (n.d.). Remote calibration of a GPS timing receiver to UTC(NIST) via the Internet. In *Proceedings of the measurement science conference, Anaheim, CA*.

Lombardi, M., Novick, A., Lopez, J., Boulanger, J.-S., & Pelletier, R. (2005). The inter-American metrology system (SIM) common-view GPS comparison network. In *Proceedings of the 2005 IEEE international frequency control symposium and exposition* (pp. 691–698). IEEE.

Lombardi, M., Novick, A., & Zhang, V. (2005). Characterizing the performance of GPS disciplined oscillators with respect to UTC(NIST). In *Proceedings of the joint IEEE international frequency symposium and precise time and time interval (PTTI) systems and applications*.

Misra, P., Pratt, M., & Burke, B. (1998). Augmentation of GPS/LAAS with GLONASS: Performance assessment. In *Proceedings of ION GPS-98*.

National Coordination Office for Space-Based Positioning, N., & Timing. (2016). Retrieved from http://www.gps.gov/ (Accessed 26.04.16).

Novick, A., Lombardi, M., Zhang, V., & Carpentier, A. (1999). A high performance multichannel time-interval counter with an integrated GPS receiver. In *Proceedings of the 31st annual precise time and time interval (PTTI) meeting* (pp. 561–568).

Pratt, M., Burke, B., & Misra, P. (1998). Single-epoch integer ambiguity resolution with GPS-GLONASS L1–L2 data. In *Proceedings of ION GPS-98*.

Seo, J., Walter, T., Chiou, T.-Y., & Enge, P. (2009). Characteristics of deep GPS signal fading due to ionospheric scintillation for aviation receiver design. *Radio Science*, *44*(1).

Sharma, K. (2012). *Fundamentals of radar, sonar and navigation engineering (with guidance)*. New Delhi: S.K. Kataria & Sons.

Sokolova, N., Borio, D., Forssell, B., & Lachapelle, G. (2010). Doppler rate measurements in standard and high sensitivity GPS receivers: Theoretical analysis, and comparison. In *Proceedings of the second international conference on indoor positioning and indoor navigation (IPIN), Zurich, Switzerland*.

St. John's College, University of Cambridge (2014). *The way to the stars: Build your own astrolabe*. Retrieved from http://www.joh.cam.ac.uk/library/library_exhibitions/schoolresources/astrolabe/build/ (Accessed 26.04.16).

Uren, J., & Price, W. (2010). *Surveying for engineers* (5th ed.). London, New York, Shanghai: Palgrave Macmillan.

Vonderohe, A. (n.d.). *Accuracies of laser rangefinders and GPS for determining distances on golf courses*. Retrieved from http://laserisbetter.com/wordpress/wp-content/uploads/2009/03/accuraciesofrangefinders-drvonderohe.pdf (Accessed 26.04.16).

Weiss, M., Zhang, V., White, J., Senior, K., Matsakis, D., Mitchell, S., ... Proia, A. (2011). Coordinating GPS calibrations among NIST, NRL, USNO, PTB, and OP. In *Joint conference of the IEEE international frequency control and the European frequency and time forum (FCS) proceedings* (pp. 1070–1075). IEEE.

Zhu, Y., Wang, Y., Forman, G., & Wei, H. (2015). Mining large-scale GPS streams for connectivity refinement of road maps. *The Computer Journal*, *58*(9), 2109–2119.

INDEX

Note: Page numbers followed by *f* indicate figures and *t* indicate tables.

Printed in the United States
By Bookmasters